A History of the Third Offset, 2014–2018

GIAN GENTILE, MICHAEL SHURKIN, ALEXANDRA T. EVANS,
MICHELLE GRISÉ, MARK HVIZDA, REBECCA JENSEN

Prepared for the Joint History and Research Office
Approved for public release; distribution unlimited

For more information on this publication, visit www.rand.org/t/RRA454-1

Library of Congress Cataloging-in-Publication Data is available for this publication.
ISBN: 978-1-9774-0626-2

Published by the RAND Corporation, Santa Monica, Calif.

Cover: NicoElNino / Adobe Stock.

Preface

This report documents the history of the Third Offset from 2014 to 2018. The Third Offset was a competitive strategy that sought to capitalize on the potential for certain technologies to offset the recent military advances of China and Russia. This history focuses on institutional efforts to effect change within the U.S. Department of Defense (DoD) and the key defense leaders who strove to bring that change to fruition. Over a four-year period starting in 2014, senior DoD leaders—above all Deputy Secretary of Defense Robert O. Work, who served in that role from mid-2014 to mid-2017—developed working groups and agencies to push the ideas of the Third Offset. Their efforts were successful in that the 2018 National Defense Strategy embraced many of the fundamental tenets of technological advances and organizational changes developed by the Third Offset. In that sense, this history provides an example of how to effect organizational and process changes in large military institutions like DoD. This research should be of interest to policymakers, historians, and scholars with an interest in organizational change, the impact of emerging technologies, and strategic competition.

The research reported here was completed in November 2020 and underwent security review with the sponsor and the Defense Office of Prepublication and Security Review before public release.

This research was sponsored by the Joint History and Research Office and conducted within the International Security and Defense Policy Center of the RAND National Security Research Division (NSRD), which operates the National Defense Research Institute (NDRI), a federally funded research and development center sponsored by the Office of the Secretary of Defense, the Joint Staff, the Unified Combatant Commands, the Navy, the Marine Corps, the defense agencies, and the defense intelligence enterprise.

For more information on the RAND International Security and Defense Policy Center, see www.rand.org/nsrd/isdp or contact the director (contact information is provided on the webpage).

Contents

Tables

Summary

The Third Offset emerged at a time of important transition within the U.S. Department of Defense (DoD). In 2014, the U.S. wars in Afghanistan and Iraq, which had started in 2001 and 2003, respectively, seemed to be winding down. At the same time, it had become clear to U.S. defense planners that for the previous two decades, when the U.S. military was concentrated on Iraq and Afghanistan, China and Russia had significantly increased their warfighting capabilities.[1] Both countries had produced long-range air defense and fires systems that made it much more difficult for the United States to project combat power in a conflict in Asia or Europe. As senior U.S. defense leaders looked at the worldwide security environment, they saw an increasingly militarily capable China and Russia and a U.S. military that had consumed its energies in two-decades-long small wars and lost its conventional warfighting edge.[2]

The aim of the Third Offset, as envisioned by former Deputy Secretary of Defense Robert O. Work, one of its key creators and advocates, was to draw on U.S. advanced technologies, such as artificial intelligence, cyber capabilities, unmanned systems, and machine learn-

---

[1] This was the perception among U.S. defense planners. This report does not consider the extent to which U.S. intelligence assessments agreed that Russian and Chinese warfighting capabilities had improved.

[2] In other words, there was a growing realization that China and Russia had increased their warfighting capabilities. At the time, however, the idea that China and Russia were adversaries or strategic competitors of the United States remained somewhat heretical. This suggests that the realization that China and Russia had increased their warfighting capabilities did not automatically translate into a widespread belief that they were now competitors of the United States. This transition took time.

ing, to name a few, to offset—or create an overmatch of—China's and Russia's increased capabilities. But for Work and others, better military technologies to counter Chinese and Russian advances, to be achieved through streamlined business processes and improved coordination between DoD and industry, were not enough. Those improved U.S. technologies needed to be combined with new organizational constructs and future warfighting concepts for the U.S. military. At the same time, there needed to be a rethinking of how U.S. foreign policy treated Russia and, especially, China. Instead of treating both countries as potential partners and focusing on strengthening economic and diplomatic ties, Work and others believed that the United States should treat China and Russia as strategic competitors and, in a crisis, even enemies. The Third Offset therefore comprised each of these three elements.

Of course, the existence of a Third Offset implies that a first and second offset preceded it. Indeed, from 2014 to 2018, proponents of the Third Offset were careful to stress what they viewed as the historical continuity of their efforts by interweaving it with the story of previous successful efforts, in the 1950s and 1970s, to deploy advanced U.S. technologies to "offset" Soviet conventional military superiority. Many DoD senior leaders noted that Work and others on his team had a deep understanding of history, which informed the development of the Third Offset between 2014 and 2018.[3] But what was distinctive about the team was not just their deep sense of historical precedent but rather that they shared an awareness of the importance of using historical narrative to promote the Third Offset within DoD.

This report focuses not so much on the First and Second Offsets that came before the Third Offset; rather, it describes the real intellectual changes that the Third Offset fostered within DoD. There are several ways to judge the success of the Third Offset. First, it may be judged in terms of the degree to which it offset Russian and Chinese

---

[3] The starting point for this history, 2014, marks the beginning of Work's tenure as deputy defense secretary. Although Work left DoD in 2017, we use 2018—the year that the National Defense Strategy (NDS), which incorporated many of the ideas identified with the Third Offset, was adopted—as the endpoint for this history.

capabilities. Second, it may be judged by the extent to which its core principles were adopted in the NDS in 2018. Third, it may be judged in terms of its role in facilitating cooperation between DoD and Silicon Valley. As we note in this report, it is still too early to say whether the Third Offset has, or ultimately will, offset Russian and Chinese advanced capabilities. According to the second and third metrics, however, the Third Offset succeeded. The Third Offset was intended not only to develop specific technologies but also to be a mechanism of change that would force DoD to start to look at current and future U.S. security problems in a different light. In that regard, it succeeded. In the end, after four years of Third Offset ideas working through and permeating DoD thinking, the proof of its influence was the 2018 NDS, which reflected many of the fundamental ideas and tenets of the Third Offset.

Acknowledgments

We would like to thank the two reviewers of this history—Gregory Daddis, the Nimitz Foundation Chair in Modern Military History at San Diego State University, and Joshua Klimas of the RAND Corporation—for their most helpful and insightful reviews. We also would like to thank RAND's Dave Ochmanek and Stephanie Young for an earlier review of a different version of this history. Our sponsor at the Joint History Office, David Crist, gave us thorough support and encouragement throughout the research and writing process and very helpful input on this history. Christine Wormuth, our program director at RAND, provided us with the resources and guidance that we needed to carry this history through to completion. Lastly, we want to thank our great support team at RAND—Nancy Pollock, Pete Ledwich, Taria Francois, and Saci Detamore—for helping us see this through from start to finish.

Abbreviations

A2/AD	anti-access/area denial
AAN	Army After Next
ACDP	Advanced Capabilities and Deterrence Panel
AI	artificial intelligence
C2	command and control
CAPE	Cost Assessment and Program Evaluation
CCMD	combatant command
CIA	Central Intelligence Agency
CJCS	Chairman of the Joint Chiefs of Staff
CNAS	Center for a New American Security
COIN	counterinsurgency
CSBA	Center for Strategic and Budgetary Assessments
DARPA	Defense Advanced Research Projects Agency
DII	Defense Innovation Initiative

DIUx	Defense Innovation Unit—Experimental
DoD	U.S. Department of Defense
DSB	Defense Science Board
FCS	Future Combat Systems
IC	Intelligence Community
ICBM	intercontinental ballistic missile
ISR	intelligence, surveillance, and reconnaissance
JICSpOC	Joint Interagency Combined Space Operations Center
LOE	line of effort
LRRDPP	Long-Range Research and Development Planning Program
MDO	multidomain operations
NATO	North Atlantic Treaty Organization
NDS	National Defense Strategy
ONA	Office of Net Assessment
OSD/P	Office of the Under Secretary of Defense/Policy
PDDNI	Principal Deputy Director of National Intelligence
PGM	precision-guided munitions
PLA	People's Liberation Army
RMA	Revolution in Military Affairs

SCO	Strategic Capabilities Office
SEAL	Sea, Air, and Land
SPR	Strategic Portfolio Review
USMC	U.S. Marine Corps
USPACOM	U.S. Pacific Command
VCJCS	Vice Chairman of the Joint Chiefs of Staff

CHAPTER ONE

Introduction

The *Third Offset* refers to an effort, led by former Deputy Secretary of Defense Robert O. Work, to draw on advanced technologies, including artificial intelligence (AI), unmanned systems, and machine learning, to offset Chinese and Russian capabilities. Beginning in 2014, when Work joined the U.S. Department of Defense (DoD), proponents of the Third Offset worked to develop the advanced technologies that would be required to create an overmatch of Chinese and Russian capabilities—and to develop closer ties between DoD and Silicon Valley tech firms.

Although the Third Offset refers primarily to this effort, it also refers to a set of ideas about the nature of the U.S. relationship with China and Russia. At the most literal level, therefore, the Third Offset refers to an initiative to replicate the so-called First and Second Offsets, in which the U.S. military ostensibly used specific technological innovations to offset certain specific advantages enjoyed by U.S. strategic competitors. In the early 1950s, during the First Offset, the United States used tactical and strategic nuclear weapons to offset the Soviet bloc's quantitative conventional advantage. In the Second Offset, from the mid-1970s through the late 1980s, the United States used a combination of technologies, including precision-guided strike and stealth, to offset once more the Warsaw Pact's numerical superiority, specifically by neutralizing the second echelon of a hypothetical invasion.[1] The Third Offset was based on the presumption that, once again, technological

[1] These earlier offsets were focused on countering the Soviet Union. They did not address U.S. policy toward China.

capabilities would provide a decisive advantage for the United States, although the objective this time was not to offset adversaries' quantitative conventional advantages but rather to respond to challenges particular to fighting the two potential strategic competitors of concern, China and Russia.[2] As for which technology or technologies were to be employed in the Third Offset, one finds contradictory views. Some Third Offset leaders, foremost among them Work, explicitly cited AI, machine learning, and autonomous vehicles. Others—and sometimes Work—were agnostic about the specific technologies to be employed so long as they were linked to operational concepts. The point was to invest in innovation with the understanding that some new capabilities would emerge and once again give the U.S. military an edge.

On another level, however, the Third Offset refers more loosely to a set of ideas. One of these ideas was the conviction that China and Russia (especially China) were in fact strategic competitors of the United States. This ran counter to what amounted to official thinking until well into the second administration of President Barack Obama (2013–2017). The corollary to this idea was the conviction that the United States needed to develop a strategy for competing with China and Russia and make that strategy the centerpiece of its national defense strategy. This meant, among other things, refocusing the military on acquiring the kinds of capabilities required to confront strategic competitors, which was something that it had not been doing for at least a decade. This included a focus on countering China's and Russia's anti-access/area denial (A2/AD) technologies, which were of particular significance because of the U.S. military's need to project forces to the theaters of operation. Another idea pertained to the relationship between DoD and industry, and DoD's ability to drive and harness

---

[2] Here, it could be argued that, in characterizing the Third Offset as linked to these earlier offsets, there is an inherent presumption that the first two offsets were successful and were therefore suitable models for the Third Offset. However, because the earlier offsets were never tested by a Warsaw Pact invasion, it is difficult to assess the extent to which they were successful. It remains unclear whether technological advantages would have enabled North Atlantic Treaty Organization (NATO) forces to successfully repel a Soviet conventional offensive. As a result, the foundational assumptions of the Third Offset were based on an understanding of history that cannot be proven.

the results of innovation. The Third Offset reflected the observation that DoD could no longer drive innovation as it had during the First and Second Offsets, given that most technological innovations were now coming from the commercial sector—especially Silicon Valley, which was not particularly interested in selling to the U.S. military. The Third Offset therefore featured a drive to find new ways to cultivate technological innovations and interact with the commercial world, including Silicon Valley.

Related to the Third Offset's ideas about technology was its "enterprise" angle. The Third Offset was animated by the idea that DoD not only needed to refocus its attentions but also had to change how it did business, especially in relation to the acquisitions process—i.e., cultivating and acquiring new technologies, absorbing innovations, and developing entirely new operating concepts to make use of them.

Lastly, there was a human side to the Third Offset. Although by no means its sole author, Work was the primary advocate of the Third Offset and the person who did the most to make it a reality. He began that process soon after he assumed his position in DoD in 2014, and, in the fall of that year, he created two new institutions, the Advanced Capabilities and Deterrence Panel (ACDP) and the so-called Breakfast Club, which were the key drivers of the Third Offset. These institutions were responsible for promoting Third Offset ideas and shepherding and integrating numerous lines of effort before winding down in 2017. In many ways, their story is synonymous with that of the Third Offset as a whole.

If one focuses exclusively on the first, literal meaning of the term, the idea of the Third Offset and its impact tends to generate skepticism among scholars and commentators. One problem is that to write about the First and Second Offsets is to, to some extent, apply post hoc terms that were adopted after the First and Second Offsets, first by a few of their architects, but then, later and more fully, by advocates of the Third Offset to make their case. This invites debate regarding whether past developments occurred the way that they are portrayed by the official narrative. Another problem is that the idea of the Third Offset, at least according to prevailing popular understandings of its role and impact, appears to be synonymous with faith that a few specific technologies, such as AI, machine learning, and autonomous vehicles, might serve as a silver bullet and change the strategic equation in the U.S. military's favor.[3] Such enthusiasms are reminiscent of buzzwords from the 1990s and early 2000s, such as Army After Next (AAN), Revolution in Military Affairs (RMA), and Transformation. It would be difficult to argue that the Third Offset, so defined, was a success.

It is possible to understand the Third Offset, and Work's considerable role in shaping defense policy, in a different light, however, if we think of the Third Offset as representing a broader set of ideas about innovation and adaptation. The Third Offset marked a change in thinking in the Pentagon and the interagency with regard to great-power competition, the relationship between DoD and industry, and how DoD conducts much of its business. Although the Third Offset might not have been the direct cause of that change, it almost certainly made a significant contribution.

It might be useful to think of the Third Offset's history of opening intellectual doors as a metaphor for the trajectory of the ideas that it encompassed. In 2014, the ideas that Work espoused and used the ACDP and Breakfast Club to promote were unpopular within DoD, notwithstanding the official imprimatur of Defense Secretary Chuck Hagel, who signed the memo that marks the official birth of the Third Offset. Roughly four years later, Work's ideas had been enshrined in the 2018 National Defense Strategy (NDS).[4] As we argue in this report, the Third Offset ended in large part because Work succeeded. Or at least he did if one defers judgment on the extent to which the Third Offset contributed to the offset of certain Chinese and Russian capabilities in the manner of the First and Second Offsets, through the development and deployment of AI, machine learning, and autonomous vehicles. Whether that has or ever will come to fruition, we cannot say. The development of these specific technologies is, we argue, somewhat less

---

[3] For an example of the belief that technology can have a transformative impact on U.S. military strategy, see Frederick W. Kagan, *Finding the Target: The Transformation of American Military Policy*, New York: Encounter Books, 2006.

[4] Jim Mattis, *Summary of the 2018 National Defense Strategy of the United States of America: Sharpening the American Military's Competitive Edge*, Washington, D.C.: U.S. Department of Defense, 2018.

important given the Third Offset's real accomplishment—opening the door to a new way of thinking about great-power competition and the relationship between DoD and industry.

Historians attempting to write the history of the Third Offset therefore must address it in all of its meanings. They have to write about the ideas and the people; they have to write at once an intellectual history and an institutional and organizational history. Given how recent these events were, we have had little choice but to rely heavily on interviews with some of the people involved, supplemented by open source documents. Ideally, we would have used more primary source documents. However, they survive primarily on hard drives and servers, making access difficult and uneven. Those that we did use were provided to us by the Joint History Office and consisted primarily of documents generated by a few individuals in the Office of the Secretary of Defense. These are mostly memos and briefings. Other documents were provided to us by individuals involved, from their private archives; these consist primarily of briefings.

Methodologically speaking, using first-person accounts for historical work presents certain problems owing to the vagaries of memory and the subjective nature of any individual's account of past events. Rather than eschew such material, however, the appropriate approach is to handle personal accounts with care. This means, among other things, presenting people's views as no more than that: people's views. We quote what people say and make clear that what we cite is someone's view, as opposed to presenting the information as fact. Where there are divergent opinions, we note them. At the same time, we are careful not to report disagreements for their own sake; we did not invite our interviewees to talk about one another beyond seeking clarification of individual roles or contributions.

It should also be stated in advance that, although we invited our interviewees to tell us what they thought they, or the Third Offset in general, accomplished, we pass on this document not as an audit. Skepticism is entirely appropriate, and the reader will find it throughout the text with the interpretive angles that we take toward our topic. However, we believe that we have treated our topic fairly and that we give credit where credit is due. Work managed to alter the course of a ship as large as DoD and to influence how people thought and spoke about several important matters, and he did so largely by convincing people with ideas and inspiring them. This is an impressive feat regardless of one's views of the Third Offset. Indeed, this history is, in part, a study of leadership and, relatedly, of institutional change.

In this report, we first explain the origins of the ideas behind the Third Offset. In Chapter Two, we examine the history of the two previous offsets, as told by Third Offset policy advocates, to convey the history of strategic thinking in DoD and the role of technology therein. The Third Offset represents, among other things, a return to a focus on peer adversaries and great-power competition after 30 years of the defense establishment perceiving no peer threats and nearly two decades of focusing on counterterrorism and the wars in Afghanistan and Iraq. Chapter Two also introduces Work and the genesis of the ideas behind the Third Offset.

Chapter Three reviews what Work did beginning in 2014, when he assumed his position as Deputy Secretary of Defense. It presents the many initiatives and organizations that became associated with the Third Offset or that Work stood up specifically to support the Third Offset. These include the ACDP and the Breakfast Club, which together constituted the key institutions of the Third Offset; the many elements associated with the Third Offset's lines of effort (LOEs); and such organizations as the Strategic Capabilities Office (SCO) and the Defense Innovation Unit—Experimental (DIUx). Chapter Three closes with the winding down of the Third Offset in 2017 and 2018. As we will discuss, the end of the Third Offset came about in part because of changes in DoD leadership, but, by that point, it had already made a significant contribution to changing how many in the national defense establishment thought and talked about peer threats, military capabilities, and technology.

CHAPTER TWO

Setting the Scene

This chapter discusses the context of the Third Offset with respect to two interrelated areas: (1) the evolution of defense planning priorities and (2) thinking about great-power competition. In many ways, the Third Offset represented a return to a way of perceiving U.S. security threats that was predominant during the Cold War, when the United States faced a particular peer-state adversary and sought to generate a credible military deterrence, which required the development of certain specific military capabilities.[1] That approach lost ground after the fall of the Berlin Wall, when the United States no longer faced any peer adversaries, and even more so after the terrorist attacks of September 11, 2001, when DoD became focused on the so-called Global War on Terror and the wars in Afghanistan and Iraq. The Third Offset represents a shift back to thinking in terms of strategic competitors.

Of course, there are important differences between the second decade of the 21st century and the Cold War, many of which have to do with an understanding of the specific challenges represented by adversaries' militaries, as well as with an appreciation for new technological innovations and changes in the relationship between government and industry. Both the continuity and the changes are captured by Work's choice of describing what he wanted to achieve as the *Third*

---

[1] It should be noted, of course, that the Cold War was not fought only in Western Europe. Although U.S. defense planners and strategists may have been focused on countering the Soviet threat against NATO in Western Europe, the Cold War also comprised a multitude of proxy wars in the Global South. See Odd Arne Westad, *The Global Cold War: Third World Interventions and the Making of Our Times*, New York: Cambridge University Press, 2005.

Offset, which he saw as both similar to and different from what he described as the First and Second Offsets.

Such terms as the *First* and *Second Offset*, it should be clear, are ahistorical. During the First Offset, which ostensibly took place during the administration of President Dwight D. Eisenhower, no one referred to it as such. A few people, including Secretary of Defense Harold Brown, referred to the Second Offset, which began during the administration of President Jimmy Carter, as an offset, albeit not in capital letters.[2]

However, the accuracy of this historical narrative, as told by Work and his colleagues, is beside the point. As former Principal Deputy Director of National Intelligence (PDDNI) Stephanie O'Sullivan put it, Work "wanted to inspire people and give them permission to think outside the lines," and the "myth" of the First and Second Offsets helped do that.[3] As the eminent military historian Sir Michael Howard observed, one of the central purposes of military history is the creation of myths that tell the story and lineage of combat units to help build cohesion.[4] Work and his colleagues similarly used a narrative of the past as a tool that enabled them to pursue their policy goals.

Still, in telling the history of the Third Offset, we appreciate that this narrative was useful for a similar reason: It provided a concise and effective way to communicate the evolution of the larger strategic context after World War II, and the Pentagon's responses, all leading up to Work's endeavors upon becoming Deputy Secretary of Defense in 2014.

---

[2] William J. Perry, "Desert Storm and Deterrence," *Foreign Affairs*, Vol. 70, No. 4, Fall 1991.

[3] Stephanie O'Sullivan, interview with RAND Corporation researchers about the Third Offset and Robert Work, Arlington, Va., September 18, 2019.

[4] Michael Howard, "The Use and Abuse of Military History," *RUSI Journal*, Vol. 107, No. 625, 1962.

The Cold War and the First Offset

During the Cold War, the obvious preoccupation of the U.S. Armed Forces was countering the threat represented by the Soviet Union and its Warsaw Pact allies. This priority informed a variety of decisions regarding what kind of military the United States would have, as well as its equipment, doctrine, and so on. Of paramount concern was countering the Soviet bloc's vast numerical advantage in conventional forces in Western Europe. At first, the United States enjoyed a nuclear monopoly, which gave it some assurance. Even so, the Soviet Union had nearly three times the number of conventional ground forces as the United States and its key allies, who defended Western Europe from a ground invasion from the east.[5] Eisenhower considered it economically unrealistic for the United States to triple its number of conventional forces in Europe, including armor and infantry divisions and artillery formations. The cost of doing so, in his view, would cripple the then-growing and vibrant U.S. economy of the mid-1950s. America's European allies, moreover, could not or would not significantly increase their own share of the burden in light of economic and political constraints.

The Eisenhower administration developed a strategy designed to offset both the Soviets' advantage in conventional troops and their nascent nuclear arsenal while balancing domestic economic considerations and NATO politics.[6] Under this strategy, the ability of U.S. forces to inflict "massive retaliatory damage by offensive striking power"—i.e., both tactical and strategic nuclear fires—would make it possible to maintain a smaller conventional presence in Europe and deter the Soviet leadership, which presumably valued their individual

---

[5] For a critical analysis of U.S. assessments of the Soviet military, see Richard A. Bitzinger, *Assessing the Conventional Balance in Europe, 1945–1975*, Santa Monica, Calif.: RAND Corporation, N-2859-FF/RC, 1989.

[6] The strategy, which was laid out in National Security Council Paper 162/2, came to be known as the *New Look*; see James S. Lay, *A Report to the National Security Council*, Washington, D.C.: National Security Council, NSC 162/2, October 30, 1953.

and national survival, from invading Western Europe.[7] This strategy of investing in nuclear technology and cultivating the ability to respond massively and quickly to an invasion with nuclear weapons was the First Offset, although the term was not used at the time.

By the end of the 1950s, Eisenhower's New Look and massive retaliation came under criticism from different parts of the U.S. national security establishment, leading to the development of new organizational concepts. The U.S. Army, for one, developed a new organizational structure for its combat divisions to fight on a nuclear battlefield. The Pentomic Division, as it was called, was pushed on the Army by its chief of staff, General Maxwell D. Taylor, to give the Army a role in a potential nuclear war with the Soviets.[8] Yet as soon as its first combat divisions transformed to the new Pentomic structure, the new organizational structure came under heavy criticism from ranks within the Army. The doctrine that went along with the new organizational structure was seen as too ambitious in its reliance on future technologies that had yet to be fielded. The Pentomic Division also drew criticism for the challenges it posed to commanders by adding additional numbers of subordinate units to command control. By the early 1960s, the Army had abandoned the Pentomic structure and reverted to more-conventional organizational schemes for its combat divisions.[9]

The other major pushback toward Eisenhower's New Look strategy came from the policy realm. As the Soviet Union fielded more-capable and more-survivable nuclear forces of its own, the U.S. threat to go nuclear in response to less-than-major aggression became less and

[7] Lay, 1953, p. 582. For more on the Eisenhower administration's internal deliberations, see David Alan Rosenberg, "The Origins of Overkill: Nuclear Weapons and American Strategy, 1945–1960," *International Security*, Vol. 7, No 4, Spring 1983. See also Marc Trachtenberg, *A Constructed Peace: The Making of the European Settlement 1945–1963*, Princeton, N.J.: Princeton University Press, 1999, pp. 159–161.

[8] For further analysis, see Ingo Trauschweizer, *Maxwell Taylor's Cold War: From Berlin to Vietnam*, Lexington, Ky.: University Press of Kentucky, 2019.

[9] A. J. Bacevich, *The Pentomic Era: The U.S. Army Between Korea and Vietnam*, Washington, D.C.: National Defense University Press, 1986; and Ingo Trauschweizer, *The Cold War U.S. Army: Building Deterrence for Limited War*, Lawrence, Kan.: University Press of Kansas, 2008.

less credible.[10] Even before he was elected President in 1960, John F. Kennedy had made a sustained critique of the New Look on the grounds that it gave a U.S. President little flexibility in responding to threats to U.S. security below the nuclear threshold.[11] Once in office, Kennedy brought Taylor back onto active duty to be the Chairman of the Joint Chiefs of Staff (CJCS). Both Kennedy and Taylor argued that a new strategy for containing the Soviet Union was necessary—one that allowed for flexibility in responding to communist aggression in areas of importance to U.S. national security. Flexible Response became Kennedy's new approach to U.S. foreign policy. With that policy in place, the United States had options to respond to Soviet aggression other than retaliating massively with nuclear weapons. In the end, however, that newfound flexibility proved to be a double-edged sword because it allowed the United States to take the first incremental steps that would lead to a major U.S. military commitment in Vietnam.[12]

The Second Offset

Over time, the value of that "first" offset was weakened by several subsequent developments. Among them was the qualitative and quan-

---

[10] Indeed, the Europeans themselves became increasingly worried about the implications of the New Look strategy, because they likely would bear the brunt of a nuclear exchange with the Soviet Union. Moreover, in a meeting of the National Security Council on January 5, 1955, Admiral Robert Carney, Eisenhower's chief of naval operations, argued that if the United States "tailored all [of its] military forces to a single concept of warfare, it would be unsound." He argued that U.S. forces "should have sufficient versatility to enable them to meet various circumstances short of general war, as well as general war itself" (S. Everett Gleason, "Memorandum of Discussion at the 230th Meeting of the National Security Council," Washington, D.C., January 5, 1955).

[11] It could be argued that this critique was overblown, because the Eisenhower administration used force below the nuclear threshold in supporting coups in Guatemala and Iran.

[12] John Lewis Gaddis, *Strategies of Containment: A Critical Appraisal of American National Security Policy During the Cold War*, New York: Oxford University Press, 1982; Maxwell D. Taylor, *The Uncertain Trumpet*, New York: Harper and Row, 1960; and Brian VanDeMark, *Road to Disaster: A New History of America's Descent into Vietnam*, New York: HarperCollins Publishers, 2018.

titative improvement of the Soviets' nuclear capabilities, as we have noted, which undermined Eisenhower's strategy of massive retaliation and prompted the development of new doctrines, such as graduated escalation, and experiments with force structure, such as the Pentomic Division, that were based on the assumption that the threat of an overwhelming strike in response to an act of aggression simply was not credible. Eventually, moreover, arms treaties limited the size of nuclear arsenals and strengthened the necessity of deterring aggression through conventional means (albeit assisted by tactical nuclear weapons). By the middle of the 1970s, it became clear that a second offset strategy would be necessary.

A declassified report from Brown to Congress in 1979, for example, declared that the Soviet strategic capability for a nuclear offense was already exceeding that of the United States; Soviet intercontinental ballistic missiles (ICBMs) and submarine-launched ballistic missiles were outpacing U.S. capabilities, and the U.S. advantage in nuclear-capable aircraft was rapidly declining.[13] More worrying was the projected acceleration of these trends, combined with a loss of confidence among U.S. allies regarding the status of assured destruction as an effective (or, more accurately, credible) deterrent. In addition, the Arab-Israeli wars of 1967 and 1973 had provided a glimpse of what a conventional war between the United States and the Soviet Union might resemble; both sides used platforms similar to those that the major powers would use in a clash in central Europe. The rapidity with which both sides mobilized, and the lethality and complexity of the fighting, showed that "the American tradition and Army orientation towards mobilization was an anachronism."[14] Rather than relying on vast potential resources and the fact that the United States was protected from potential adversaries by an ocean, which had served the United States well in two world wars and Korea, a new level of techno-

---

[13] Analysts have debated the extent to which the Soviet Union had achieved strategic parity with the United States, although there has been no rigorous evaluation of the validity of Brown's claims in the declassified report.

[14] Richard Lock-Pullan, "'An Inward Looking Time': The United States Army, 1973–1976," *Journal of Military History*, Vol. 67, No. 2, April 2003, p. 498.

logical modernization would be necessary to avoid the casualty levels of the Yom Kippur War, which reached 50 percent for armored units.[15]

As a result, a new group of senior defense leaders determined that a new strategy was needed to offset the Soviet Union's conventional superiority and ward off the threat of an armored assault across central Europe. By leveraging its technological advantages to develop new force multiplier capabilities, the United States could mitigate the Warsaw Pact's quantitative advantage. As William Perry, who helped design the initiative as Under Secretary of Defense for Research and Engineering from 1977 to 1981 and built on it as Deputy Secretary of Defense from 1993 to 1994, clarified in a later article, the offset's purpose was not only to "build better weapon systems than those of the Soviet Union."[16] Rather, new capabilities could give outnumbered U.S. forces "a significant competitive advantage over their opposing counterparts by supporting them on the battlefield with newly developed equipment that multiplied their combat effectiveness."[17] As one of the leading members of this group, Andrew Marshall, noted, an additional benefit of this approach was that the pressure to match U.S. modernization efforts might strain the Soviet economy and force the Kremlin to make uncomfortable choices. If the United States could "induce Soviet costs to rise" by driving the Soviets to try to match U.S. mod-

---

[15] For illustrative examples of the lessons that U.S. and allied strategists drew from the 1973 war, see J. R. Transue, *Assessments of the Weapons and Tactics Used in the October 1973 Middle East War*, Arlington, Va.: Institute for Defense Analyses, Weapons Systems Evaluation Group Report 249, October 1974; and Lawrence Whetten and Michael Johnson, "Military Lessons of the Yom Kippur War," *The World Today*, Vol. 30, No. 3, March 1974.

[16] Perry, 1991, pp. 68–69.

[17] Perry, 1991, pp. 68–69. Defense Secretary Harold Brown offered a similar assessment in a 1981 report to Congress, in which he noted:

> Technology can be a force multiplier, a resource that can be used to help offset numerical advantages of an adversary. Superior technology is one very effective way to balance military capabilities other than matching an adversary tank-for-tank or soldier-for-soldier. (Harold Brown, *Department of Defense Annual Report Fiscal Year 1982*, Washington, D.C.: U.S. Department of Defense, January 19, 1981, p. x)

See also Robert Martinage, *Toward a New Offset Strategy: Exploiting U.S. Long-Term Advantages to Restore U.S. Global Power Projection Capability*, Washington, D.C.: Center for Strategic and Budgetary Assessments, 2014.

ernization efforts, the United States could "complicate Soviet problems in maintaining its competitive position."[18]

This effort to devise a new strategy to compensate for the loss of U.S. nuclear superiority was linked closely with the 1976 articulation of the Active Defense doctrine.[19] Relying primarily on existing technologies, and at a period when night-vision technology and precision-guided munitions (PGMs) were still embryonic, Active Defense battle emphasized targeting a concentration of enemy offensive forces on a narrow front to strike deep behind the Soviet front and destroy the staging upon which Soviet second-echelon capability depended.[20] If the Soviets' depth permitted them to lose the first battle of a war and regroup to fight effectively, Active Defense assumed that NATO had no similar luxury.[21] Besides, a less offensive defense-in-depth strategy would all but guarantee the destruction of West Germany, which

---

[18] A. W. Marshall, *Long-Term Competition with the Soviets: A Framework for Strategic Analysis*, Santa Monica, Calif.: RAND Corporation, R-862-PR, 1972, p. 33. The concept is described further in A. W. Marshall and James Roche, "Strategy for Competing with the Soviets in the Military Sector of the Continuing Political-Military Competition," unpublished Department of Defense memorandum, 1976. See also Albert Wohlstetter's 1974 article accusing the Central Intelligence Agency (CIA) of underestimating Soviet military capabilities (Albert Wohlstetter, "Is There a Strategic Arms Race?" *Foreign Policy*, No. 15, Summer 1974). This view was also raised in connection with the 1975 Team B exercise and the creation of the Committee on the Present Danger, as well as by other bipartisan nongovernmental organizations and prominent defense intellectuals and former policymakers who sought to raise the alarm about Soviet advances and critique détente.

[19] From 1977 to 1980, constant-dollar military outlays rose from $216.4 billion to $229.4 billion and included significant investments in such programs as the MX missile and the B-2 stealth bomber, ICBM force upgrades, and improvements in nuclear command and control (C2) systems. This growth continued under the next administration; from 1980 to 1986, the constant-dollar defense budget grew by over 40 percent. See Hal Brands, *Making the Unipolar Moment: U.S. Foreign Policy and the Rise of the Post–Cold War Order*, Ithaca, N.Y.: Cornell University Press, 2016, pp. 38–39, 76.

[20] The move to active defense was a doctrinal shift; an *offset* can be described as a subtype of doctrinal shift.

[21] John L. Romjue, *From Active Defense to AirLand Battle: The Development of Army Doctrine 1973–1982*, Fort Monroe, Va.: U.S. Army Training and Doctrine Command, June 1984, pp. 3–16.

NATO ostensibly aimed to protect; Germans understandably found such a strategy hard to accept.

Closely associated with this Second Offset strategy was the Long-Range Research and Development Planning Program (LRRDPP). Beginning in the early 1970s, under the auspices of the Advanced Research and Projects Agency (later the Defense Advanced Research Projects Agency, [DARPA]), the program prioritized the development of precision weapons, based on the premise that conventional but extremely accurate munitions could cause damage sufficient to obviate the need for nuclear weapons in most instances.[22] Indeed, architects of the offset argued that the new generation of conventional weapons could "shift the competition to a technological area where we [the United States] have a fundamental long-term advantage," as Perry testified to Congress in 1978. "Precision guided weapons . . . have the potential of revolutionizing warfare," he argued, and could "greatly enhance our ability to deter war without having to compete tank for tank, missile for missile with the Soviet Union."[23]

Also associated with Second Offset was the AirLand Battle Doctrine, first published in 1982. It built on improved technology that greatly enhanced surveillance, target acquisition, C2, and precision targeting to offer a model of combined arms warfare that integrated tactical air support and relied upon cooperation with other U.S. and

---

[22] Benjamin M. Jensen, "The Role of Ideas in Defense Planning: Revisiting the Revolution in Military Affairs," *Defence Studies*, Vol. 18, No. 3, July 2018, p. 308. DARPA also encouraged the development of other defining offset systems, including anti-armor weapons, space-based infrared sensors, and stealth. See Robert Tomes, "The Cold War Offset Strategy: Origins and Relevance," War on the Rocks, November 6, 2014. This statement reflects a pervasive influence of techno-optimism—according to which technological innovation can solve all manner of societal problems, including challenges facing the U.S. defense establishment—on American life. Margaret O'Mara, "The Church of Techno-Optimism," *New York Times*, September 28, 2019.

[23] Testimony of William Perry in U.S. House of Representatives, *Hearings on Military Posture and H.R. 10929: Department of Defense Authorization for Appropriations for Fiscal Year 1979: Hearing Before the Committee on Armed Services, House of Representatives, Ninety-Fifth Congress, Second Session*, HASC No. 95-56, Part 3, Book 1, Washington, D.C., U.S. Government Printing Office, 1978, p. 1049.

allied services.[24] The doctrine advocated for the adoption of a maneuver warfare ethos and sought to use the force-multiplying effect of coordinating precision attacks to counter Soviet echelonment, in which successive masses of ground forces prepared to exploit any advances made by the first echelon. Improved air defenses and air assault, as well as the use of tactical air capabilities for more than close air support, illustrated the effort of the AirLand Battle Doctrine to apply new technologies and concepts of fighting to overcome the Soviet Union's nuclear parity and conventional advantage.[25]

Finally, the Second Offset influenced the large modernization program that began under Carter and received a boost under President Ronald Reagan, yielding what often is referred to as the "Reagan Buildup." This brought a bevy of new weapon systems, replacing the Vietnam-era kit with the M-1 Abrams tank; the M-2 Bradley infantry fighting vehicle; fourth-generation fighter aircraft; and an array of missiles and rockets, many of them guided, such as cruise missiles; as well as radar systems, information networks, and stealth technologies, among others.

The architects of the Second Offset did not envision hardware and technology as ends in themselves. Sophisticated equipment and systems required conceptual innovation to produce a superior way of fighting. Also, under the Carter administration, the financial resources available were compatible with existing defense budgets, and, although applied research into battlefield technology was part of the overall strategy, at no point was future or hypothetical technology central to either Active Defense or AirLand Battle. The former relied upon the technology of the 1960s and early 1970s, while the latter integrated proven breakthroughs in precision targeting; C2; and intelligence, surveillance, and reconnaissance (ISR). The conceptual dimension of the Second Offset was also clear, and key: Technology that the Soviet Union did not have and might not be able to afford would be used

---

[24] Robert A. Gessert, "The AirLand Battle and NATO's New Doctrinal Debate," *RUSI Journal*, Vol. 129, No. 2, 1984.

[25] Manfred R. Hamm, "The AirLand Battle Doctrine: NATO Strategy and Arms Control in Europe," *Comparative Strategy*, Vol. 7, No. 3, 1988.

to enable a more effective doctrine designed to offset traditional and numerical Soviet strengths.

The U.S. military never tested its Second Offset investments and doctrine against Soviet forces, but it did deploy them against the Iraqi military during Operation Desert Storm in 1991.[26] The war, in which a U.S.-led coalition defeated the world's fourth-largest army within six weeks and with minimal losses, seemed to confirm that the offset had driven a "revolutionary advance in [U.S.] military capability," as Perry wrote shortly after the war.[27] Although later research has emphasized the Iraqi Army's structural weakness, the unity of the multinational effort, the quality of its leadership, and the efficacy and depth of logistical support systems, champions of the new technology and related doctrine argued that the war had validated their theory that improvements in sensing, targeting, and precision had produced a transformative result.[28] Operational approaches and systems "engineered and acquired in the late 1970s through the late 1980s" had made U.S. victory "inevitable and our historically small loss of life probable," Vice CJCS (VCJCS) Admiral Bill Owens said.[29] Some in the military even discerned the outlines of what they referred to at the time as a "Revolution in Military Affairs."

---

[26] Stephen Biddle, *Military Power: Explaining Victory and Defeat in Modern Battle*, Princeton, N.J.: Princeton University Press, 2004.

[27] Perry, 1991, p. 68. In his essay on the Gulf War, Perry emphasizes one illustrative example of such criticism: James Fallows, *National Defense*, New York: Random House, 1981.

[28] Several analysts have questioned the assumption that Desert Storm validated the theory of investing in Second Offset technologies and doctrine, noting that the Iraqi military was far weaker than the peer adversaries that the capabilities were designed to confront. See Andrew J. Bacevich, "'Splendid Little War': America's Persian Gulf Adventure Ten Years On," in Andrew J. Bacevich and Efraim Inbar, eds., *The Gulf War of 1991 Reconsidered*, Abingdon, UK: Routledge, 2003, pp. 149–161; Stephen Biddle, "The Gulf War Debate Redux: Why Skill and Technology Are the Right Answer," *International Security*, Vol. 22, No. 2, Fall 1997; Thomas G. Mahnken and Barry D. Watts, "What the Gulf War Can (and Cannot) Tell Us About the Future of Warfare," *International Security*, Vol. 22, No. 2, Fall 1997; and Keith L. Shimko, *The Iraq Wars and America's Military Revolution*, New York: Cambridge University Press, 2010.

[29] As quoted in Tomes, 2014.

The arguments about RMA precipitated a succession of initiatives and modernization programs in the 1990s and early 2000s, some vast in scale and ambition.[30] These included AAN and Transformation, Future Combat Systems (FCS), and the F-22 and F-35 aircraft, among others, all of which were premised on assumptions about the ostensibly paradigm-shifting virtues of "information dominance" and networking. One feature of that era that distinguishes it from the Second Offset, however, was that, whereas the Second Offset came in response to specific battlefield challenges that had emerged in the Cold War in the 1970s, by the 1990s, there was no specific challenge and the U.S. government lacked a comparable strategic focus. In a sense, DoD was searching for a new raison d'être, while the nation's security strategy had coalesced around the conviction that engagement with China and with America's erstwhile enemy, Russia, was an imperative, and while the country also worked to strengthen multilateral institutions. As for modernization, in the absence of clear "pacing" threats, defense leaders tended to look at these advanced technologies as giving them a capabilities-based DoD that could respond to a wide range of threats, not a singular threat, such as the Soviet Union during the Cold War.

DoD thinking shifted even more profoundly in the years following the terrorist attacks of September 11, 2001, which had the effect of modernization—especially with respect to conventional capabilities—receiving less emphasis. By 2004, the U.S. military—primarily the Army and the U.S. Marine Corps (USMC)—found itself heavily engaged in two counterinsurgency (COIN) wars in Afghanistan and Iraq, wars for which it was not well prepared. Consequently, DoD largely shifted its gaze from conventional capabilities and major combat operations against regional adversaries to focus on COIN and stability operations.[31] Transformation-inspired modernization programs faltered. One notorious example

[30] Although the extent to which this process of technological development was appropriately synchronized with warfighting realities and defense acquisition processes can be debated, such a debate is beyond the scope of this report.

[31] Although this reflects the shift within DoD as a whole, the extent to which this applied equally across all services is unclear.

is FCS, which Secretary of Defense Robert Gates cancelled in 2009. There were many reasons for the cancellation of FCS, but it is generally understood to be the case that DoD leadership had become less appreciative of expensive modernization efforts that did not align with the military's immediate requirements in Afghanistan and Iraq. Gates also terminated the U.S. Air Force's program to develop a new stealth bomber and truncated the F-22 fighter buy from 770 planes to 280.

Obama-Era Thinking on Global Conflict: 2009–2014

The Obama administration initially expressed little interest in the prospect of commencing a major modernization effort, at least one geared toward enhancing capabilities required for a fight with a peer adversary.[32] In a series of high-profile statements during its first year, the new administration embraced a strategy of "comprehensive engagement" premised on repairing U.S. alliances; rehabilitating the national reputation of the United States; rebuilding the domestic foundations of national power; and facilitating greater multilateral cooperation on complex global issues, such as climate change.[33] As Derek Chollet, who served as Assistant Secretary of Defense for International Security Affairs from 2012 to 2015, later wrote, "Obama took issue with how the Washington wisdom defined 'strength.'"[34] Through treaties, alli-

---

[32] Indeed, after President Obama signed a military policy bill in October 2009, the *New York Times* reported that the Obama administration had "[set] a tone of greater restraint than the Pentagon had seen in many years" and "trim[med] more weapons systems than any president had in decades" (Christopher Drew, "Victory for Obama over Military Lobby," *New York Times*, October 28, 2009). The Obama administration had sought to achieve more–fiscally responsible defense policies.

[33] Hal Brands, *American Grand Strategy in the Age of Trump*, Washington, D.C.: Brookings Institution Press, 2018, pp. 57–58; and Derek Chollet, *The Long Game: How Obama Defied Washington and Redefined America's Role in the World*, New York: PublicAffairs, 2016, pp. 43, 47, 53–54. For a sample of Obama's campaign rhetoric, see Barack Obama, "A New Strategy for a New World," speech, Washington, D.C., July 15, 2008.

[34] Chollet, 2016, pp. 42, 45. In his 2006 memoir, Obama claimed that the end of the Cold War and the integration of Germany and Japan into the international economy "effectively eliminated the threats of great-power conflicts inside the free world" competition and state

ances, and agreements, the President argued, the United States could "help create a shared set of norms to shape state behavior, widening the circle of the global order led by the United States" and regenerating U.S. leadership at lower cost.

Indeed, the administration's conviction that U.S. interests were best secured through cooperation, not competition, rested on the assumption that sustained diplomatic engagement would discourage destabilizing behavior and bring China and Russia into the liberal democratic fold.[35] Although the wars in Afghanistan and Iraq had shaken policymakers' confidence in other arenas, their thinking remained closely tied to the post–Cold War consensus of gradual liberalization, which stipulated that robust diplomatic engagement would encourage economic liberalization and that the international expansion of prosperity would promote peace.

This belief shaped the administration's views of China. Senior officials recognized that the country was a rising power and expressed concern over its "long-term, comprehensive military modernization" efforts,[36] but the dominant post–Cold War logic of liberalization suggested that Communist Party leadership in Beijing, much like their counterparts in Moscow, could not maintain an authoritarian regime indefinitely. The challenge for the United Sates, therefore, was to discourage Chinese aggression while expanding bilateral cooperation on areas of mutual interest, and, in the process, to encourage the country's gradual opening of its political and economic systems. As VCJCS Admiral James "Sandy" Winnefeld, Jr., commented to reporters, "Don't push China . . . we can all get along."[37]

---

expansion. Looking to China and Russia, he argued that the world's "most powerful nations . . . are largely committed to a common set of international rules" (Barack Obama, *The Audacity of Hope: Thoughts on Reclaiming the American Dream*, New York: Crown Publishing Group, Crown Publishers, 2006, p. 305).

[35] This assumption is somewhat evocative of President Richard Nixon's approach to China; Margaret MacMillan, *Nixon and Mao: The Week That Changed the World*, New York: Random House, 2007.

[36] DoD, *Quadrennial Defense Review Report*, Washington, D.C., February 2010, p. 31.

[37] Jeffrey A. Bader, *Obama and China's Rise: An Insider's Account of America's Asia Strategy*, Washington, D.C.: Brookings Institution Press, 2012, p. 146.

To senior policymakers, the need to work with China was also a "practical necessity."[38] The Obama administration confronted several daunting challenges in the region—including territorial disputes over the South China Sea; advances in North Korean nuclearization; and economic and political uncertainty in Thailand, the Philippines, Indonesia, and Malaysia—that required Beijing's cooperation. These areas of common interest provided opportunities for the administration, which endeavored to foster a stable and cooperative relationship with the Chinese leadership through frequent, substantial, and high-level strategic dialogues on a range of economic, political, and security issues.[39] To reassure and expand the ranks of U.S. partners in the region, Washington also sought to invest in regional multilateral institutions, to reaffirm the U.S. commitment to the freedom of the seas, to invest in regional allies and security partners, and to rebalance U.S. military assets to the Asia-Pacific arena.[40] There was a military dimension to this strategy, but the administration "always stressed that the rebalance to Asia was not just about ships and planes."[41] Shortly after leaving office in 2011, Jeffrey Bader, Obama's principal adviser on China, wrote:

> A sound China strategy should rest on three pillars: (1) a welcoming approach to China's emergence, influence, and legitimate expanded role; (2) a resolve to see that its rise is consistent with international norms and law; and (3) an endeavor to shape the Asia-Pacific environment to ensure that China's rise is stabilizing rather than disruptive.[42]

---

[38] Chollet, 2016, p. 57.

[39] Bader, 2012, p. 143.

[40] Bader, 2012, pp. 4, 143–144; Brands, 2018, p. 60; Chollet, 2016, pp. 55–56.

[41] Chollet, 2016, p. 56.

[42] Bader, 2012, p. 7. The National Security Strategy (NSS) acknowledged that the rebalance to Asia should encompass not only the rebalance of military assets but also increased cooperation with regional allies to respond to the threats posed by "terrorism, climate change, international piracy, epidemics, and cybersecurity, while achieving balanced growth and human rights" (Barack Obama, *National Security Strategy*, Washington, D.C.: The White House, May 2010, p. 42).

The administration applied similar logic to Russia, which, unlike China, was regarded as a declining power. The administration maintained the long-held U.S. objective of a democratic and stable Russia at peace with its neighbors, and it hoped to bring that objective about through, among other things, cooperation on areas of shared interest, such as nuclear security, trade, Afghanistan, and North Korea. Although factions of his administration pushed for a reevaluation of the Russian threat to global stability, the President maintained that the Russian leadership recognized U.S. dominance and that their belligerent behavior was a reflection of fundamental weaknesses.[43] As Obama explained to one interviewer, Russian President Vladimir Putin remained "constantly interested in being seen as our peer and as working with us, because he's not completely stupid. He understands that Russia's overall position in the world is significantly diminished."[44]

By Obama's reelection in 2012, however, the administration's confidence in the possibility of cooperation with Russia had already begun to dim. The halting progress of negotiations on critical political and economic issues illustrated that the areas of common interest between the countries were narrower—and the disagreements on core issues, such as human rights, trade rules, and freedom of the seas, far deeper—than senior officials had expected.[45] It was also apparent that the regional military balance was continuing to shift in China's favor, despite the Pentagon's efforts to redirect U.S. assets to the Asia-Pacific. The administration had recognized the ramifications of China's development of sophisticated A2/AD weapons in its 2010 Quadrennial Defense Review, which directed the U.S. Navy and the Air Force to develop the AirSea Battle concept, expand long-range strike capabilities, increase resilience of U.S. forward bases, address threats to space-

[43] Chollet, 2016, pp. 64–65, 163.

[44] Jeffrey Goldberg, "The Obama Doctrine," *The Atlantic*, April 2016.

[45] Chollet, 2016, p. 58; and David E. Sanger, *Confront and Conceal: Obama's Secret Wars and Surprising Use of American Power*, New York: Crown Publishing Group, Broadway Paperbacks, 2012, p. 377.

based assets, and take other steps to counter Chinese advances.[46] But in the early 2010s, while the U.S. military struggled to compensate for budget cuts and to balance competing demands in Iraq and Afghanistan, the growing investments in modernization of the People's Liberation Army (PLA) were already "altering the security balance in the Asia Pacific [and] challenging decades of U.S. military preeminence in the region," a 2013 congressional review determined.[47] Senior officials monitored China's expanding investment in anti-ship cruise and ballistic missiles, anti-satellite technology, diesel and nuclear submarines, and other advanced capabilities that "target[ed] [U.S.] vulnerabilities, not our strengths," as Gates wrote.[48]

This created the perception that the Chinese (as well as the Russians) had stolen a march on the United States by gaining a technological advantage as Washington was looking elsewhere.[49] The U.S. military had become so consumed with Iraq and Afghanistan that it had "stuck its head in the sand" and ignored other potential adversaries, Winnefeld later noted.[50] Beijing had improved its integrated air

---

[46] DoD, 2010, pp. 31–34, 59–60. In addition to its ongoing military modernization efforts, China was engaging in increasingly assertive behavior in the South and East China Seas. See Mark E. Manyin, Stephen Dagget, Ben Dolven, Susan V. Lawrence, Michael F. Martin, Ronald O'Rourke, and Bruce Vaughn, *Pivot to the Pacific? The Obama Administration's "Rebalancing" Toward Asia*, Washington, D.C.: Congressional Research Service, R42448, March 28, 2012.

[47] U.S.-China Economic and Security Review Commission, *2013 Report to Congress of the U.S.-China Economic and Security Review Commission*, Washington, D.C.: U.S. Government Printing Office, November 2013, p. 17; see also Brands, 2018, p. 173. On China's military modernization, see Eric Heginbotham, Michael Nixon, Forrest E. Morgan, Jacob L. Heim, Jeff Hagen, Sheng Li, Jeffrey Engstrom, Martin C. Libicki, Paul DeLuca, David A. Shlapak, David R. Frelinger, Burgess Laird, Kyle Brady, and Lyle J. Morris, *The U.S.-China Military Scorecard: Forces, Geography, and the Evolving Balance of Power, 1996–2017*, Santa Monica, Calif.: RAND Corporation, RR-392-AF, 2015.

[48] Robert M. Gates, *Duty: Memoirs of a Secretary at War*, New York: Alfred A. Knopf, 2014, p. 528.

[49] The degree to which the Chinese and Russians had attained a technological advantage can be debated, but such a debate is beyond the scope of this report.

[50] James Winnefeld, Jr., interview with RAND Corporation researchers about the Third Offset, McLean, Va., July 15, 2019.

defense system to protect China proper and to threaten any potentially hostile aircraft hundreds of miles from its coasts. Russia also had made significant improvements in similar areas. The end result was to put a serious dent in the ability of the United States to do what it had always done since the end of the Cold War: project military power around the globe without interference from an adversary.

To make matters worse, there was a series of "tit-for-tat nastiness,"[51] including a cyberattack on the defense secretary's office and incidents of harassment of U.S. ships transiting through the South China Sea, which compounded senior officials' suspicions that Beijing had interpreted U.S. leaders' efforts at rapprochement as evidence of U.S. weakness.[52] As one senior official later recounted, it was clear that "the relationship is fundamentally one of competition."[53] At the same time, this view was held by a minority.

The debate over China's intentions intensified in early 2011, when the PLA tested its new J-20 stealth fighter jet, designed to avoid U.S. detection and close the technology gap, during Gates's visit to Beijing.[54] The move, one senior aide later told reporter David Sanger, was "a giant screw-you to Gates and Obama."[55] But the J-20 test development also drew senior officials' attention to a second, deeper shift in the relationship between technological innovation and national power. Photos of the J-20's cockpit revealed unusual similarities to the U.S.

---

[51] Sanger, 2012, p. 329.

[52] Sanger, 2012, pp. 330, 392–393.

[53] Robert O. Work, interview with RAND Corporation researchers about the Third Offset, Arlington, Va., June 24, 2019a. Reflecting on the Obama administration's shifting views toward China in a later interview, Work noted that "*Competition* was a word that . . . didn't convey what we were trying to do" during the first administration (quoted in Uri Friedman, "The New Concept Everyone in Washington Is Talking About," *The Atlantic*, August 6, 2019; emphasis in original). "By the end . . . the administration just said, 'Hey, China is truly a competitor, and we need to hedge against future bad behavior.'"

[54] Elisabeth Bumiller and Michael Wines, "Test of Stealth Fighter Clouds Gates Visit to China," *New York Times*, January 11, 2011.

[55] Sanger, 2012, p. 390. Reflecting on the incident in his memoirs, Gates quoted an aide's assessment at the time: "This is about as big a [pushback] as you can get" (Gates, 2014, p. 527).

F-22 advanced fighter jet, deepening suspicions that the Chinese aircraft's development had been informed by data stolen from U.S. contractors in 2009.[56]

The problem of stealing data from U.S. contractors highlighted another aspect of competition with China related to the relationship between government and technology that would become an important feature of Third Offset thinking. Since the end of World War II, U.S. dominance in science and technology had been maintained through investment in government laboratories and contracted research and development centers, producing a system in which the enforcement of traditional regulatory mechanisms, such as U.S. export controls and other restrictions on dual-use technologies, allowed for tight control of any sensitive technologies. However, with the acceleration of technological progress over the period from the 1980s through the 2000s, private-sector corporations, such as Google and Apple, had begun to outpace the work of their public and semipublic counterparts—and to outgrow a dated regulatory system ill-equipped for their futuristic technologies. The bottom line was that, in many key fields, the private sector, not the Pentagon, was driving innovation. This complicated the acquisition process and also made it significantly more difficult to maintain any kind of technological edge, given China's access to American high-tech industries and its role in the global supply chain. Matt Turpin, who worked for Winnefeld and later served as assistant to VCJCS General Paul J. Selva, explained the problem in terms of a toy from the 1980s known as *Teddy Ruxpin*, a bear that possessed a rudimentary ability to speak.[57] According to Turpin, export of the toy

---

[56] U.S.-China Economic and Security Review Commission, *2012 Report to Congress of the U.S.-China Economic and Security Review Commission*, Washington, D.C.: U.S. Government Printing Office, November 2012. In 2009, cyber intruders linked to China were reported to have successfully targeted vulnerabilities in contractors' networks to access and copy several terabytes of design and electronics systems data related to the Pentagon's Joint Strike Fighter project; Siobhan Gorman, August Cole, and Yochi Dreazen, "Computer Spies Breach Fighter-Jet Project," *Wall Street Journal*, April 21, 2009. See also Bill Gertz, "Report: China's Military Is Growing Super Powerful by Stealing America's Defense Secrets (Like the F-35)," *National Interest*, December 8, 2016.

[57] Matt Turpin, interview with RAND Corporation researchers about the Third Offset, Washington, D.C., June 26, 2019.

to the Soviet Union had been blocked because some of the electronics inside could be reverse engineered and used to advance the Soviets' own capabilities. By the mid-2010s, however, such a thing could not be done, according to Turpin, because the Chinese had arranged it so that they would be supplying the components in question.

The dangers were fourfold. First, as a U.S. government report noted, the complexity of technical systems and the increasing fragmentation of U.S. supply chains presented China and other strategic competitors with a greater number of opportunities for corporate espionage, theft, and subversion. That China, long recognized as "the world's 'biggest thief' of American intellectual property,"[58] possessed sophisticated cyber capabilities only compounded the defensive challenges.[59] Second, the relocation of U.S. high-technology, manufacturing, and research and development facilities to China during the 1990s and early 2000s had left U.S. corporations vulnerable to Beijing's overt and covert efforts to capture U.S. intellectual property. Confronted with the option of complying with Chinese government demands or losing access to factories and the booming Chinese market, U.S. corporations had proven willing to share once–closely held trade secrets. Third, private technology corporations seemed to prefer commercial rather than DoD contracts and might willingly promote the spread of dual-use technologies in foreign markets.[60] This suggested that DoD had to think creatively about how to work with private technology companies to obtain needed technologies. Finally, Chinese state-owned

---

[58] U.S.-China Economic and Security Review Commission, *2014 Report to Congress of the U.S.-China Economic and Security Review Commission*, Washington, D.C.: U.S. Government Printing Office, November 2014, p. 92.

[59] See, for instance, the case of Yu Long, a permanent U.S. resident and Chinese citizen who pled guilty to charges that he stole, sold, and transported documents related to a sensitive military program at United Technologies (Office of Public Affairs, U.S. Department of Justice, "Chinese National Admits to Stealing Sensitive Military Program Documents from United Technologies," press release, last updated December 22, 2016). For an example of the heightened attention to Chinese cyber capabilities, see Office of the National Counterintelligence Executive, *Foreign Spies Stealing US Economic Secrets in Cyberspace: Report to Congress on Foreign Economic Collection and Industrial Espionage, 2009–2011*, Washington, D.C.: Office of the Director of National Intelligence, October 2011.

[60] Turpin, 2019.

enterprises could infiltrate the U.S. supply chain by investing in or purchasing relevant companies to gather information; assess U.S. capabilities; or challenge the system's integrity through the introduction of counterfeit, substandard, or intentionally subverted components—a prospect confirmed in 2012, when a Senate Armed Services Committee investigation identified China as "the dominant source country for counterfeit electronic parts" that already had infiltrated a variety of military systems.[61]

In short, the United States now needed to mitigate the effects of new Chinese capabilities, strengthen control over sensitive technologies, deter or deny further Chinese theft and subversion, and identify and extricate Chinese points of influence within the supply chain. There was little time to catch up. Congressional investigations conducted in 2012 confirmed that current U.S. export controls were inadequate and that Chinese state-owned enterprises had already infiltrated the U.S. national security and high-technology sectors.[62] Chinese President Xi Jinping's ascension in 2012 and the subsequent intensification of the country's modernization efforts only lent additional credibility to the emerging narrative that China was intending to surpass the United States.[63]

Frustrated with the scale of cybertheft, and under pressure from Congress to tighten procurement regulations and contracting procedures to limit foreign interference, the Obama administration began to

---

[61] U.S.-China Economic and Security Review Commission, 2012, p. 7. The lack of transparency surrounding the government's role in Chinese companies compounds the challenge. Although the Chinese government of the Communist Party might seek to control a company's activities through direct ownership or control, government and party entities also might exercise indirect influence through a corporation's board or management team.

[62] For illustrations of how the Chinese challenge was perceived during this period, see the 2009–2014 U.S.-China Economic and Security Review Commission reports to Congress, which can be found at U.S.-China Economic and Security Review Commission, "Annual Reports," webpage, undated.

[63] Sanger, 2012, pp. 330, 392–393. On Xi's policy shift, see Andrew Chubb, "Xi Jinping and China's Maritime Policy," Brookings Institution, January 22, 2019; and U.S.-China Economic and Security Review Commission, 2013.

roll out new guidelines for critical-sector corporations in 2012.[64] "This is billions of dollars of combat advantage for China. They've just saved themselves 25 years of research and development. It's nuts." Another senior official put it more bluntly: "This idea of pulling ahead with an offset was nearly impossible if the Chinese were in the car with us."[65]

At the same time as U.S. officials were confronting the extent of China's military modernization, new concerns arose over Russia's intensifying aggression. The administration's efforts to "reset" the U.S.-Russian relationship produced a series of early diplomatic successes, including an agreement on United Nations sanctions on Iran in 2010, the signing of the New Strategic Arms Reduction Treaty in 2011, the establishment of the Northern Distribution Network in Afghanistan, and Russia's entry into the World Trade Organization.[66] To the White House's frustration, however, these demonstrations of U.S. goodwill did not prevent the Russians from demonstrating a pattern of worsening behavior. In 2012, Putin's reclamation of the presidency ushered in a period of heightened tensions, worsened by the Russian leader's circulation of conspiracy theories of NATO aggression and U.S. efforts to foment protests against his regime. Russia's military modernization efforts continued apace, and Putin authorized a series of provocative large-scale military exercises that featured a war against a

---

[64] Eric A. Fischer, Edward C. Liu, John W. Rollins, and Catherine A. Theohary, *The 2013 Cybersecurity Executive Order: Overview and Considerations for Congress*, Washington, D.C.: Congressional Research Service, R42984, December 15, 2014. Reflecting the heightened anxiety over the integrity of the U.S. supply chain, Congress passed legislation in 2012 that required DoD to "report . . . on the extent to which its current procurement regulations and contracting procedures allow it to exclude the acquisition of any foreign-produced equipment from any department system where there is concern as to the potential impact of cyber vulnerabilities" (U.S.-China Economic and Security Review Commission, 2012, p. 23). Soon after, Pentagon and White House officials publicly attributed a series of cyberespionage incidents to the Chinese government and military and called upon Beijing to end its corporate espionage programs. See Ernesto Londoño, "Pentagon: Chinese Government, Military Behind Cyberspying," *Washington Post*, May 6, 2013; and Ellen Nakashima, "U.S. Publicly Calls on China to Stop Commercial Cyber-Espionage, Theft of Trade Secrets," *Washington Post*, March 11, 2013.

[65] Turpin, 2019.

[66] Chollet, 2016, p. 160.

NATO-like force and a simulated nuclear strike on Warsaw.[67] Then, in February 2014, the Russian president ordered the invasion of Ukraine and the annexation of Crimea, precipitating the most significant crisis in Europe since the fall of the Berlin Wall. The war did not change the Obama administration's conviction that Russia was a declining power, but it affirmed voices within the administration who had long cautioned that the United States was confronting a new period of great-power competition.[68] The stage was set for what would be known as the Third Offset.

Meet Robert Work

The history of the Third Offset simply cannot be written without placing Work at center stage. Although far from being the sole author of the Third Offset, he was the most prominent senior defense leader to not only author the ideas behind it but also push it aggressively in DoD. Work also contributed many of the intellectual concepts that underpinned the Third Offset while he elaborated and articulated the narrative of the First and Second Offsets and best made the case for the Third.

Work's introduction to military technology occurred at the Naval Reserve Officers Training Corps at the University of Illinois, where he was commissioned as a second lieutenant in the artillery branch of the USMC. Unlike the Air Force and the Navy, the USMC is not known as a technology- or equipment-based force. Artillery is the USMC's most technical and technology-based branch, however, and it provided Work with a professional understanding of military technologies. Over his 27 years of active commissioned service, he commanded from platoon to battalion level, before retiring as a colonel in 2001.

---

[67] Chollet, 2016, pp. 159–161. This new era of great-power competition consisted of a blend of conventional threats, similar to those encountered during the Cold War, and new technological threats.

[68] Chollet, 2016, p. 162; Friedman, 2019; and Robert O. Work, "Remarks by Defense Deputy Secretary Robert Work at the CNAS Inaugural National Security Forum," transcript, Center for a New American Security, December 14, 2015b.

After leaving the military, Work joined the Center for Strategic and Budgetary Assessments (CSBA), a Washington, D.C., think tank known for its vocal advocacy of the RMA.[69] The new position placed Work within a network of defense luminaries, such as Andrew Krepinevich and Andrew Marshall. Over the next nine years, first as a Senior Fellow and later as Vice President for Strategic Studies, he published extensively on the importance of maintaining the U.S. military's conventional technological advantage by developing and acquiring new systems and improving the capabilities of existing systems.[70] While at CSBA, Work also familiarized himself with the organization's history of extensive collaboration, between 1995 and 2000, with the Office of Net Assessment (ONA). During this period, CSBA and ONA had developed the "Future Warfare 20XX" series of wargames, which tested RMA-associated capabilities against a "large peer competitor" explicitly "modeled on a rising China."[71] Among the capabilities attributed to the hypothetical adversary were advanced air and sea area-denial technologies, which the games obliged Blue teams to confront.[72] The games also included space and cyberspace as battlefields, and they placed considerable emphasis on unmanned air, ground, surface, and underwater vehicles with at least some degree of autonomy.[73]

In 2009, Obama nominated Work to be Under Secretary of the Navy, the service's second-highest position. In that role, Work dis-

---

[69] For an illustrative example of CSBA's work on military modernization during this period, see Michael G. Vickers and Robert C. Martinage, *The Revolution in War*, Washington, D.C.: Center for Strategic and Budgetary Assessments, December 2004.

[70] See, for instance, Thomas P. Ehrhard and Robert O. Work, *The Unmanned Combat Air System Carrier Demonstration Program: A New Dawn for Naval Aviation?* Washington, D.C.: Center for Strategic and Budgetary Assessments, May 10, 2007; and Thomas P. Ehrhard and Robert O. Work, *Range, Persistence, Stealth, and Networking: The Case for a Carrier-Based Unmanned Combat Air System*, Washington, D.C.: Center for Strategic and Budgetary Assessments, 2008.

[71] Michael G. Vickers and Robert C. Martinage, *Future Warfare 20xx Wargame Series: Lessons Learned Report*, Washington, D.C.: Center for Strategic and Budgetary Assessments, December 2001, p. 1.

[72] Vickers and Martinage, 2001, pp. 9–10, 13–14.

[73] Vickers and Martinage, 2001, p. 15.

played an interest in many of the policies that would later coalesce into the Third Offset. Work understood that China and Russia were close to achieving technological parity with the United States,[74] and, as a result, he questioned former Defense Secretary Donald Rumsfeld's longstanding optimism that the United States held a clear technological edge over its opponents and pushed continuously for greater investment in unmanned and advanced computing systems.[75] He also talked about great-power competition.[76]

In 2013, after four years with the Navy, Work left the Pentagon to become chief executive officer of the Center for a New American Security (CNAS), a bipartisan Washington think tank. Work would serve only one year in this role before being called back to the Pentagon to serve as Defense Secretary Chuck Hagel's deputy. Nonetheless, the brief stint was a formative period for Work, who later credited his time at CNAS for giving him necessary physical and intellectual space from DoD. Work's time at CNAS confirmed his view from his previous stint as Under Secretary of the Navy that Chinese and Russian technological advances undermined the foundation of U.S. conventional deterrence toward present and future adversaries. Something needed to be done, he believed. With Shawn Brimley, a former director of strategic planning in the Obama administration's National Security Council and CNAS's founding director of studies, in 2014, Work published a series of reports refining a new offset strategy to restore the qualitative

[74] Work, 2019a.

[75] Robert O. Work, "The Coming Naval Century," *Proceedings*, Vol. 138, No. 5, May 2012; and Robert O. Work and F. G. Hoffman, "Hitting the Beach in the 21st Century," *Proceedings*, Vol. 136, No. 11, November 2010. In a reflection of his priorities, Work raised the issue of military modernization during his confirmation hearings, stating his belief that the United States was "on the cusp of a revolution in unmanned technologies" (U.S. Senate, *Nominations Before the Senate Armed Services Committee, First Session, 111th Congress: Hearings Before the Committee on Armed Services*, Washington, D.C.: U.S. Government Printing Office, 2010).

[76] Jim McCarthy, interview with RAND Corporation researchers about the Third Offset, Arlington, Va., November 13, 2019.

edge of the United States.[77] One of those studies notably bears the title *20YY: Preparing for War in the Robotic Age*, suggesting a meaningful continuity between it and the *20XX* wargames that CSBA and ONA conducted during the 1990s. The last section of the 20YY report consists of a call to arms that, in many ways, stands as a summary of the Third Offset:

> Since the end of the Cold War, the United States military has enjoyed a virtual monopoly in the guided munitions-battle network regime. . . .
>
> Now, however, as the United States' ability to project power and to dominate force-on-force encounters begins to erode as more and more opponents become able to effectively employ guided weapons, defense planners must begin to shift their gaze from the current war-fighting regime to the coming one dominated by proliferated sensors, electric weapons, and ubiquitous unmanned and autonomous systems in all operating domains. Unfortunately, as they do so, there is a very real danger that today's environment . . . will make it challenging to spur and sustain the thinking, development of new operational concepts, research, experimentation and investments needed to prepare today's U.S. military for the demands of the 20YY future. . . .
>
> The United States must overcome this challenge. If it hopes to maintain its technological superiority, the U.S. armed forces must begin to conceptualize how a maturing guided munitions-battle network regime and advances in technologies driven primarily by the civilian sector may coalesce and combine in ways that could spark a new military-technical revolution. It cannot afford to defer the time, thinking and investments needed to prepare for warfare in the Age of Robotics. . . . To a degree that U.S. force planners are simply not accustomed to, other global actors are in a position to make significant headway toward a highly robotic

[77] Robert O. Work and Shawn Brimley, *20YY: Preparing for War in the Robotic Age*, Washington, D.C.: Center for a New American Security, January 2014.

war-fighting future in ways that could outpace the much bigger and slow-moving U.S. defense bureaucracy.

The United States cannot allow this to happen.[78]

By the time Work returned to the Pentagon on May 1, 2014, as Deputy Secretary of Defense (see Table 2.1), he had a relatively clear idea of what he wanted to achieve.

The Idea of the Third Offset

The atmosphere in Washington had begun to shift during the year Work spent at CNAS. The PLA's land reclamation efforts in the South China Sea and increased investment in military modernization had empowered voices within the Obama administration who now publicly warned of the potential dangers arising from China's rapid ascendance. The first hint of this eventual shift occurred prior to Work's tenure at CNAS, in 2012, when then–Deputy Secretary of Defense Ashton Carter stood up SCO specifically to attempt to find near-term solutions for countering new Chinese military capabilities.[79] By the time Work rejoined DoD, Winnefeld had also become increasingly concerned with Chinese capabilities and was doing what he could to push DoD to focus on the problem, making him a willing partner of Work

Table 2.1
Deputy Secretaries of Defense in the 2010s

Deputy Secretary of Defense	Term
Ashton Carter	October 6, 2011–December 3, 2013
Robert O. Work	May 1, 2014–July 14, 2017
Patrick M. Shanahan	July 19, 2017–January 1, 2019

[78] Work and Brimley, 2014, p. 36.

[79] Greg Grant, interview with RAND Corporation researchers about Grant's experiences while working on the Third Offset, telephone, September 27, 2019.

(see Table 2.2).[80] By 2014, according to former PDDNI O'Sullivan, parts of the Intelligence Community (IC) also were coming around to the idea of viewing China and Russia as strategic competitors; however, there was not yet a consensus, and both views were still included in reports.[81] According to Greg Grant, who began working at the Pentagon as Gates's speechwriter and later worked for Work, around 2014 all of the combatant commands (CCMDs), but especially U.S. Pacific Command (USPACOM), suggested that they did not have the basic tools they needed to fight the Chinese.[82]

Then, in March 2014, Russia invaded Ukraine, a move heralded in Washington as evidence of Putin's ambitions to restore the country's great-power status. Once taboo, discussion of a new era of great-power competition now circulated within the administration. In a speech at the Council on Foreign Relations, Work noted:

> I left as the Under Secretary of the Navy, and I thought at that time that the challenges facing the Department of Defense were quite daunting. Now . . . as the daily headlines attest, we face even greater geopolitical challenges than I would have even dreamed of

Table 2.2
Vice Chairmen of the Joint Chiefs of Staff During the Third Offset

Vice Chairman	Term
Admiral James A. Winnefeld, Jr.	August 4, 2011–July 31, 2015
General Paul J. Selva	July 31, 2015–July 31, 2019

[80] Grant, 2019.

[81] O'Sullivan, 2019. There was a lack of consensus because, although it was apparent that Chinese capabilities were increasing, it was not yet clear what China intended to achieve with these capabilities.

[82] Grant, 2019. Of note, USPACOM was the official command title until 2018, when it became USINDOPACOM (U.S. Indo-Pacific Command).

> . . . and a far more frustrating and challenging budget environment here at home.[83]

This did not mean that by 2014 everyone was convinced that a new era of great-power competition had arrived. Work and others complained that, until the end of the Obama administration, the White House would not accept language describing China as a potential adversary, because, at the time, the administration was trying to establish a friendlier relationship with China. Still, tectonic plates were shifting. However, they were not shifting fast enough for Work.

To Work, the new strategic landscape generated an urgent need to regain the U.S. historic technological overmatch against potential adversaries. Chinese and Russian investments in PGMs, battle networks, and air defense over the previous 20 years had allowed both countries to approach parity with the United States. Worse still, China's modernization efforts appeared specifically designed to mitigate the advantages of the United States. The operational results were vastly improved A2/AD capabilities that could challenge U.S. force projection in the region. Work worried that, unless the United States invested in aerial and naval unmanned systems, AI, computer-assisted human operation systems, and AI-enabled battle networks, among other technological innovations, it would lose the ability to deter its adversaries and defend its interests abroad. In an important qualification about the importance of technology, Work also noted that "technology is never, never the final answer"; defense institutions need "to be able to incorporate those technologies into new operational and organizational constructs."[84]

Work credits Paul G. Kaminski (who served as, among other things, Special Assistant to Under Secretary of Defense Perry during the Second Offset, in which Kaminski played a lead role in the development of stealth technology) with bringing the term *offset* to his atten-

---

[83] Robert O. Work, "A New Global Posture for a New Era," transcript of speech delivered to Council on Foreign Relations, Washington, D.C., September 30, 2014.

[84] Work, 2019a. See also Paul J. Selva and Robert O. Work, unpublished interview about the Third Offset, October 24, 2016; and Work, 2015b.

tion and providing the basic idea of a "third" offset as part of a lineage that began with the "first" and "second" offsets.[85] In a video on YouTube dated January 22, 2014,[86] which Work says he watched, Kaminski tells the story of how the United States used technology to "offset" the Soviets' advantages.[87] In his presentation, Kaminski did not refer to a "first" or "second" offset. Instead, he spoke of an "offset strategy," which the United States used on two occasions. The first time was in the early 1950s, in the years immediately following the end of World War II, when the United States made the choice of using nuclear weapons to offset the Soviet military's vastly greater size. The second time, from the mid-1970s through the end of the 1980s, was when Soviet gains with respect to their nuclear capabilities theoretically had diminished the value of the "first" offset. The idea of the "second" offset was to re-tip the scales in the favor of the United States by developing technologies that would significantly improve Western conventional forces, notwithstanding their numerical disadvantage. Work said that he subsequently spoke with Kaminski about the presentation and the idea of offsets, which gave him the narrative "hook" he required for his ideas.[88]

What Kaminski referred to as the second instance of the United States using an "offset strategy" held particular attraction for Work, in part because, according to Grant, it took place in a context that was roughly analogous to the situation in 2014.[89] In the 1970s, the U.S. military was at a low point. It had just emerged from the Vietnam War and was slowly waking up to the idea that it had missed an entire generation of modernization, allowing the Soviets to gain considerable ground. Similarly, in 2014, the U.S. military had just spent (or squandered, from a modernization point of view) a decade engaged in COIN

---

[85] Work, 2019a.

[86] TheIHMC, "Paul Kaminski: STEALTH—An Insider's Perspective," video, YouTube, January 22, 2014.

[87] Work, 2019a.

[88] Work, 2019a.

[89] Grant, 2019.

and counterterrorism campaigns.[90] Work, shortly after learning of the White House's intention to nominate him as deputy defense secretary and inspired by Kaminski's presentation, met with some of the leaders of the Second Offset, including former Secretary of Defense Perry, director of the ONA Andrew Marshall, and Kaminski himself. These defense luminaries confirmed Work's concern that the United States would soon lose the technological superiority that it had enjoyed since the end of World War II.

Work chose to brand his initiative the *Third Offset* in part to cast his ideas in terms of continuity, but also because the Second Offset strongly informed how Work intended to achieve it. For example, according to Grant, Work learned from the Second Offset that such a profound change had to be a top-down initiative from senior leadership.[91] He also learned that there had to be a focus on a specific set of military challenges, which is something that arguably distinguishes all three offsets from the RMA-linked initiatives. He wanted to resurrect the LRRDPP, which he understood to have been a key player in the Second Offset. In addition, Grant attributed Work's insight into the importance of small wins to his study of the Second Offset. Perry, Grant explained, understood that he could not simply order the Air Force to drop everything in favor of guided munitions. The services had large and expensive inventories and already-established acquisition programs, with commitments stretching years into the future. Perry recognized that the better approach was to start small and improve the technology, working incrementally until the services came around to sharing the same vision. Work, according to Grant, embraced this idea, which we will discuss later in reference to the accomplishments of SCO and DIUx. This incremental approach was also a reason, Grant asserted, for Work's emphasis on wargames. As Grant explained, Work believed that, through wargames, participants would come to understand the need for and the potential of new technologies on their own.

---

[90] This is not to suggest that defense research and development came to a standstill as a result of the wars in Iraq and Afghanistan. Rather, the requirements of this period were less tied to the development and acquisition of specific advanced technologies.

[91] Grant, 2019.

The result, in Work's view, would be a chipping away at conventional thinking within DoD that would allow new and innovative technologies and concepts to emerge within DoD in dealing with adversaries, such as China.[92]

From June to November 2014, Work intensely studied the Russian and Chinese military challenges to the United States. The glaring point of departure was that China and Russia had obtained parity in Second Offset technologies, such as PGMs and accurate long-range ground-based fires. The trick, therefore, was to discern the right technologies that would allow the United States to move from a point of parity with China and Russia to a position of advantage.

In this regard, Work's thinking was indelibly shaped by the Defense Science Board's (DSB's) *Summer Study on Autonomy*.[93] The study convinced Work that autonomy would be one of the key components—if not the key component—of any effort to offset Chinese and Russian strengths and provide U.S. combat systems with greater range and dispersal capabilities. Similarly, AI could be used to augment critical warfighting systems, such as C2, surveillance and reconnaissance, and targeting systems, for speedier effects against an adversary. Sustainment and regeneration also could be infused with autonomy enabled by AI, giving the United States a decisive advantage in future combat. The theory underpinning the Third Offset was that autonomous systems enabled by AI would allow U.S. battle command networks to gain an "operational advantage" over strategic competitors, such as China and Russia, which had shifted their theory of victory from destruction of combat systems to "battling the network."[94] Work argued that understanding the battlespace "better than the adversary" also would allow for "more rapid decision making and application of more discriminate

---

[92] Grant, 2019.

[93] The DSB's *Summer Study on Autonomy* established that autonomous technologies had the potential to mitigate many of the operational challenges facing DoD. It provided recommendations for accelerating DoD's adoption of autonomous capabilities, including increasing interaction between DoD and nontraditional research and development communities; DSB, *Summer Study on Autonomy*, Washington, D.C.: U.S. Department of Defense, June 2016.

[94] Work, 2019a.

effects faster." AI, therefore, was a key enabler for these autonomous systems to function in the high-end combat that Work envisioned.[95]

With the basic idea for a Third Offset in place, Work organized a series of meetings and dinners with staff members, such as Frank Kendall and Mike Vickers, and outside experts, among them Stephen Walt, John Mearsheimer, and Eliot Cohen. The former, he said, focused primarily on program deficiencies and desired advanced capabilities. The latter focused on the strategic implications of a return to great-power competition.[96] Work described both in terms of doing "due diligence in developing the framework of the Third Offset."

Work and his aides, Grant and Ylli Bajraktari, also developed a broader framework to implement the initiative. In June 2014, they devised five LOEs that combined technological innovation with broader institutional reforms.[97] Although Work identified "strategy" as the first LOE, it was more like a policy recommendation for treating China and Russia not as potential partners but as current strategic competitors and potential adversaries. The second LOE, which was developed by Work and his team during the second half of June, was "operational concepts." To take advantage of offsetting technological innovation, the Third Offset needed revised operational constructs from all of the U.S. military services. The third LOE concentrated on encouraging innovations. Work reestablished the LRRDPP, which was directly associated with the Second Offset, to identify and invest in next-generation concepts and technologies.[98] To establish priorities and minimize waste, Work stressed the importance of assessing feasibility. The fourth LOE was wargaming, and therefore would use wargaming

---

[95] The related concept of network-centric warfare, introduced by DoD in the 1990s, suggests that an information advantage, to be achieved via information technologies and computer networking, can be translated into a competitive advantage. The Third Offset reflected a similar effort to achieve a competitive advantage through the development and implementation of advanced technologies, although it was not limited to information technologies.

[96] Robert O. Work, email to RAND Corporation researchers about the Third Offset, September 28, 2019b.

[97] Work, 2019a.

[98] Claudette Roulo, "DoD Seeks Next-Generation Technologies, Kendall Says," U.S. Department of Defense, October 7, 2014.

tools to assess the potential use or value of each technology or concept. Work had been a proponent of using wargames as an analytical tool since his early days at CSBA. Finally, in the fifth LOE, Work stressed the importance of information management to thread the needle between "reveal[ing] capabilities for deterrence and conceal[ing] capabilities for warfighting advantage."[99]

If the basic concepts behind the Third Offset were ready by fall 2014, Work still needed to build support for the initiative within the White House, DoD, and Congress. One challenge was to convince the Obama administration that China and Russia were strategic competitors to be confronted, not potential partners to be cultivated. By this time, views about China were beginning to evolve, although the White House still shied away from the idea of great-power competition. Senior leaders in the Pentagon were more receptive to the argument, but they were preoccupied with maintaining the readiness of their forces for current operations and had paid insufficient attention, Work believed, to the future challenge of regaining a technological advantage over the Chinese and the Russians.[100]

---

[99] Ylli Bajraktari, interview with RAND Corporation researchers about Bajraktari's experiences while working on the Third Offset, Arlington, Va., July 17, 2019; and Winnefeld, 2019.

[100] Bajraktari, 2019.

CHAPTER THREE

Making It Happen: Work's Internal Initiative

The previous chapters identified and discussed the various ideas at the heart of the Third Offset, their genesis, and their evolution. However, the Third Offset also produced tangible changes in the form of several organizations charged with promoting the Third Offset and spurring and guiding activities along specific LOEs. Some of these organizations were stood up for the express purpose of supporting the Third Offset; others predated the Third Offset and were drawn into it. Some continue to exist, while others ended after a relatively brief existence.

This chapter focuses on these organizations and their efforts. Table 3.1 lists the main organizations that were associated with the Third Offset and provides a brief description of their purpose and activities. In this chapter, we pay particular attention to two entities that Work established, which together represent the institutional heart of the Third Offset: the ACDP and the so-called Breakfast Club. This chapter closes with the publication of the 2018 NDS and the roughly concomitant demise of the ACDP and the Breakfast Club. We argue that the NDS represents something of a victory for the Third Offset because it marked the embrace of what originally had been minority views within DoD and the interagency. The inclusion of these views in the new NDS meant that Work's initiative had achieved one of its central aims.

Table 3.1
Organizations Associated with the Third Offset

Organization	Purpose and Activities
Defense Innovation Initiative (DII)	Invest in the development of innovative technologies.
ACDP	Coordinate efforts across DoD to promote the Third Offset and its objectives.
Breakfast Club	Support the efforts of the ACDP by bringing together working-level representatives from across DoD; draft working documents; set the agenda for the ACDP.
DIUx	Offer contracts to tech companies to develop new technologies as needed to fulfill Third Offset goals.
SCO	Repurpose existing technologies to fulfill Third Offset goals.
Cost Assessment and Program Evaluation	Conduct Strategic Portfolio Reviews (SPRs) to support the development of new capabilities.
LRRDPP	Cultivate high-end technologies that might help the United States once more widen its technological lead over potential adversaries.

Mission Launch: The Defense Innovation Initiative

As we have discussed, the Third Offset was, in many ways, conceptually in motion well before May 2014, when Work joined DoD as deputy defense secretary. He hit the ground running, working to advance his ideas and lend them institutional substance. Among his priorities were enlisting Winnefeld in his efforts and getting Hagel on board (see Table 3.2). Working alongside Winnefeld was important because Winnefeld had significant experience and was able to provide Work with invaluable assistance in pushing through various initiatives, according to one source.[1] As for Hagel, Work made his case in a classified briefing and was thereby able to present Hagel with his vision for how to move forward.[2] According to Kathryn Harris, who was Hagel's special assistant at the time, this was the immediate impetus for two

[1] Bajraktari, 2019.

[2] Kathryn Harris, telephone interview with RAND Corporation researchers about Harris's experiences while working on the Third Offset, June 19, 2019.

Table 3.2
Secretaries of Defense During the Third Offset

Secretary of Defense	Term
Chuck Hagel	February 27, 2013–February 17, 2015
Ashton Carter	February 17, 2015–January 20, 2017
James Mattis	January 20, 2017–January 1, 2019

steps that Hagel took on November 15, 2014, which should be viewed as the start date of the Third Offset. First, he released an official memo announcing the creation of the DII, which, for a time, served as the official name for the Third Offset. Second, that same day, Hagel delivered a keynote address at the Reagan National Defense Forum that introduced the initiative and provided more details about the thinking behind it. (Interestingly, Work, in an interview for this study, said that in hindsight he thinks he would have been better served had he stuck with the name *DII*, or even possibly the *LRRDPP*, because it might have generated fewer antibodies against which he had to battle.)[3]

The memo announced the DII, which it introduced in terms of the need to establish "a broad, Department-wide initiative to pursue innovative ways to sustain and advance our military superiority for the 21st century and improve business operations throughout the Department."[4] The effort, he wrote, would be overseen by Work. As for the speech, it is particularly noteworthy as a concise expression of some of the key talking points and themes that defined the Third Offset. These include anxiety over the loss of the U.S. military's technological edge and the argument that while the United States had been focused on "grinding stability operations," Russia and China were "heavily investing in military modernization programs to blunt our military's technological edge . . . ."[5] One also finds a clear focus on preserving

---

[3] Work, 2019a.

[4] Chuck Hagel, U.S. Secretary of Defense, "The Defense Innovation Initiative," memorandum to U.S. Department of Defense staff, Washington, D.C., November 15, 2014b, p. 1.

[5] Chuck Hagel, "Reagan National Defense Forum Keynote," speech, Simi Valley, Calif., November 15, 2014a.

the U.S. military's ability to project power in light of the threat posed by potential adversaries' investment in technologies that would "limit our freedom of maneuver"[6]—i.e., A2/AD. The answer, according to Hagel, was to invest in innovation. Referring to the First and Second Offsets (although without using those terms), Hagel argued that what mattered were not the specific technologies developed in the 1950s and 1970s. On the contrary, "the critical innovation was to apply and combine these new systems and technologies with new strategic operational concepts, in ways that enable the American military to avoid matching an adversary 'tank-for-tank or soldier-for-soldier.'"

Hagel's argument highlights a contradiction in the Third Offset. Sometimes, its proponents had a clear focus on specific technologies, prominent examples being AI and machine learning. Often, however, proponents of the Third Offset were what one might call agnostic regarding specific technologies. For example, Hagel, at least in this instance, was not identifying any particular technology. For him, the point was the development of new technologies and new ways of operating them in combination, with the assumption that one way or another these new technologies would, once again, deliver an offset.

Indeed, Hagel announced the DII, which, he said, "we expect to develop into a game-changing third 'offset' strategy."[7] The idea, he explained, was not only to invest in new technology to regain America's edge but also to "change the way we innovate, operate, and do business." In other words, Hagel had in mind the broad range of ideas associated with the Third Offset, as opposed to the narrower definition of technologies that together would offset some specific advantage enjoyed by strategic competitors. Hagel provided some details about how he intended to implement the DII. One idea was to establish a new LRRDPP. Hagel also announced that Work would lead an "Advanced Capability and Deterrent Panel," the ACDP, to drive the DII forward. This panel, he explained, would "integrate DoD's senior leadership across the entire enterprise: its policies and intelligence com-

[6] Hagel, 2014a.

[7] Hagel, 2014a.

munities; the armed services; the Joint Chiefs of Staff; and research, development, and acquisition authorities."

The Advanced Capabilities and Deterrence Panel

Another institution that was central to the work of the Third Offset was the ACDP. The primary function of the ACDP was to encourage and coordinate efforts to advance the Third Offset's broad agenda; the Breakfast Club, which we discuss in detail later, did the same, only at more of a working level. ACDP meetings involved a broad range of people. They included representatives of all four services and the various regional CCMDs. Usually, the total amounted to 15 to 20 people. They met quarterly, and usually in the defense secretary's conference room on the E-ring of the Pentagon.[8]

One way to describe the ACDP's work is that it pushed various DoD and other relevant organizations to focus on Third Offset–related concerns. Another function of the ACDP was to integrate numerous parallel activities that otherwise might not converge.[9] According to Bajraktari, "The key of the ACDP was tying civilians, the IC, and the military together to move forward with Third Offset initiatives."[10] According to Grant, the ACDP was important because it brought together senior DoD leadership to get them to focus on specific problems that they now faced.[11]

The Intelligence Community and the Advanced Capabilities and Deterrence Panel

One of the participants in the ACDP was the PDDNI, which reflected an understanding that the IC both would benefit from Third Offset initiatives to develop new technologies and would support the Third

---

8 Christine Wormuth, interview with RAND Corporation researchers about the Third Offset, Arlington, Va., June 14, 2019.

9 Bajraktari, 2019.

10 Bajraktari, 2019.

11 Grant, 2019.

Offset by providing analytical support regarding, for example, assessments of other countries' technological capabilities. Former PDDNI O'Sullivan, who served as ACDP tri-chair and regularly attended ACDP meetings, explained in an interview for this study that IC cooperation came naturally because the IC, to a considerable extent, was already thinking along similar lines and doing much of what Work wanted DoD to do.[12] For example, according to O'Sullivan, there was a strong current—if not a consensus—in the IC regarding China's growing assertiveness. Interestingly, O'Sullivan argued that the value of the ACDP, for the IC, had to do with its ability to give focus to DoD's "demand signal."[13] To put it another way, the ACDP gave DoD a focus, which made it significantly easier for the IC to support DoD.

The CIA, in particular, had long since concluded that the truly innovative technological work was coming from the private sector, especially Silicon Valley. At the end of the 1990s, in recognition of this fact, the CIA had chartered In-Q-Tel, a nonprofit venture capital firm that supports the requirements of the IC. In-Q-Tel became a direct inspiration for one of the more important initiatives of the Third Offset, DIUx, which we will discuss later. In other words, according to O'Sullivan, Work did not need to convince the IC of anything—the IC was already on board. It follows that the IC's support to the Third Offset, and to the ACDP in particular, tended to fall into two categories: first, providing analytical support to work regarding Chinese and Russian capabilities and intentions, and second, providing analytical support to technological initiatives.

Advanced Capabilities and Deterrence Panel Pathfinder Projects: Special Program Missile Defeat, Joint Interagency Combined Space Operations Center, and the Algorithmic Warfare Cross-Functional Team (Project Maven)

Work's desire to demonstrate the promise of the Third Offset led him to place three "pathfinder," or demonstration, projects under the supervision of the ACDP. The first was the Special Program Missile Defeat

[12] O'Sullivan, 2019.

[13] O'Sullivan, 2019.

program. This was an effort to counter North Korea's burgeoning ICBM threat by using AI and machine learning to improve analysts' ability to exploit rapid imagery provided by spy satellites.[14]

The second was the Joint Interagency Combined Space Operations Center (JICSpOC), intended primarily to promote information-sharing on space operations between the military and the IC.[15] Work, in an address at the JICSpOC in September 2016, said that it was designed "to perform battle management and command and control of the space constellation under threat of attack" and was the "first step in the third offset to start to readdress and to extend our margin of operational superiority."[16] In April 2017, JICSpOC was renamed the National Space Defense Center. Again, a trial program launched under the aegis of the Third Offset had become a more or less permanent institution.

The third project—one that is often cited as a success—was Project Maven, formally known as the Algorithmic Warfare Cross-Functional Team.[17] The idea, announced by Work on April 26, 2017, was to demonstrate how AI and machine learning could be exploited. In this particular instance, the idea was to "turn the enormous volume of data available to DoD into actionable intelligence and insights at speed," basically by quickly processing vast amounts of drone footage to support anti–Islamic State operations in Iraq and Syria.[18]

Ironically, even though Project Maven has improved DoD's use of AI for a wide range of weapon and information systems, it has also led to a rift between DoD and some leading companies in the civilian

---

[14] A more in-depth discussion of these capabilities would exceed the classification of this report.

[15] Arthur D. Simons Center for Interagency Cooperation, "Interagency Space Center Makes Strides," March 17, 2016a.

[16] Arthur D. Simons Center for Interagency Cooperation, "JICSpOC Creates More Seamless Coordination Across Government," September 30, 2016b.

[17] Paul McLeary, "Pentagon's Big AI Program, Maven, Already Hunts Data in Middle East, Africa," *Breaking Defense*, May 1, 2018.

[18] Robert O. Work, U.S. Deputy Secretary of Defense, "Establishment of an Algorithmic Warfare Cross-Functional Team (Project Maven)," memorandum to U.S. Department of Defense staff, Washington, D.C., April 26, 2017.

sector who are at the forefront of the development of AI. One of the key aspects of the Third Offset was the realization that, for advanced military technologies, such as AI, the real innovation would come from the civilian sector and not from large U.S. defense contractors, such as Raytheon and General Dynamics. For the Third Offset to proceed, according to its founders, including Work, DoD had to acknowledge that civilian tech companies would have the lead in developing its cutting-edge technologies. The irony of Project Maven in this regard is that, in cooperating with Google, a civilian tech company, it created a mini rebellion among rank-and-file Google employees who were philosophically opposed to Google being a part of the "business of war." Google's reluctance to participate further in Project Maven was a blow to DoD in relying on civilian tech companies to further U.S. defense technologies.[19]

Responses to the Advanced Capabilities and Deterrence Panel: From Skepticism to Support

Participants in the ACDP have described an important evolution in the attitudes of many of its attendees. Those interviewed for this study said that many, if not most, participants (perhaps excluding O'Sullivan) initially greeted the initiative with skepticism. Given that the ACDP was new, both in its nature and its mission, participants were not sure what to expect at first. Everyone showed up, but mainly because the deputy defense secretary and the VCJCS were leading the meeting.[20] Nonetheless, the reception was cold. One person noted that "there were a lot of leaders in the Department who didn't buy into the idea [of the Third Offset] to begin with . . . and they figured they could just wait this out."[21] This reflects widespread skepticism throughout DoD. As some noted, among them Selva, many people at the time thought that

---

[19] Scott Shane and Daisuke Wakabayashi, "'The Business of War': Google Employees Protest Work for the Pentagon," *New York Times*, April 4, 2018.

[20] Bajraktari, 2019.

[21] Harris, 2019.

they had heard it all before, albeit under the names *Transformation* or *Revolution in Military Affairs*.[22]

Over time, however, people became less skeptical and more supportive of the ideas of the Third Offset as a starting point to begin to think differently about innovation inside DoD relative to emerging threats, such as China and its technological innovations over the preceding three decades. One participant, Christine Wormuth, confessed to dreading ACDP meetings at first, comparing them to being told to eat her vegetables. "But then I came to see it was important," she explained.[23] Another source recalls a meeting in March 2015 at which Winnefeld addressed the skepticism head on, telling those present something to the effect that he had heard that many intended to simply drag their heels and bide their time, but that this was a mistake. They needed to get on board.[24] Selva said that the rise of the Islamic State helped, because people saw what the Islamic State was doing at the "low end" of the technological spectrum, such as using unmanned aerial vehicles and melding data to target strikes, which made a real impression on them regarding what might be possible.[25] It also helped, according to another participant, that at one point Secretary of Defense Ashton Carter came to the ACDP, thereby underscoring the importance of the new initiative.[26]

Significantly, some of our sources pointed to particularly strong resistance from the services, who were accused of, among other things, being overly fixated on capacity rather than modernized capabilities: They wanted to preserve their force structure and feared that technological investments of the kind suggested by the Third Offset ultimately would translate into fewer brigades, fewer plane squadrons, and

[22] Paul J. Selva, telephone interview with RAND Corporation researchers about the Third Offset, November 18, 2019.

[23] Wormuth, 2019.

[24] Harris, 2019.

[25] Selva, 2019.

[26] Bajraktari, 2019.

fewer ships. As Selva put it, "It is fair to say that Work was thinking that the services were too focused on capacity."[27]

Jim McCarthy, a retired surface warfare officer who was involved with the Third Offset as part of his duties as the deputy to the Navy's Deputy Chief of Naval Operations for Integration of Capabilities and Resources, defended at least the Navy. He argued that the Navy did not contest the basic premises of the Third Offset or the imperative to invest in modernized capabilities. Nonetheless, the Navy still needed capacity, and it had to work with the fleet that it had, as well as with extant procurement programs.[28] Sustaining those left little room for redesigning the fleet or substantive modernization, unless the Navy was willing to "stop doing something," meaning to renounce certain missions or give up certain capabilities in the way that Britain's Royal Navy, for example, more or less gave up naval aviation for a time to help pay for new carriers and F-35s. That said, the U.S. Navy was otherwise already on a path that converged with the Third Offset, especially with respect to developing sensor technologies, long-range missiles, unmanned vessels, dispersed operations, and other capabilities.[29] The Third Offset had the effect of formalizing the processes associated with the development and acquisition of these, and other, technologies.

The Breakfast Club

The Breakfast Club supported the efforts of the ACDP by bringing together working-level representatives of mostly the same organizations represented by the ACDP. Instead of three-stars and their civilian equivalents, who populated the ACDP, the Breakfast Club consisted of "empowered people with access to the principals," or, more specifically, well-placed O5s and O6s (lieutenant colonels and colonels) and their

[27] Selva, 2019.

[28] McCarthy, 2019.

[29] For example, see John Keller, "Navy Interested in New Computing and Sensor Technologies for Shipboard and Submarine Sonar," *Military & Aerospace Electronics*, July 10, 2017.

civilian peers.[30] There are two explanations for the origin of the name, both of which could be true. There was, apparently, a group by that name run by Secretary of Defense Brown, who is associated with the Second Offset.[31] It also seems to be the case that, in Pentagon culture, "Breakfast Club" is what one calls a working group that meets early in the morning. In this case, the Breakfast Club met twice a month on Wednesdays at 8:30 a.m. in the deputy defense secretary's conference room.[32] Reportedly, breakfast was not served—not even coffee. Sometimes, Work made an appearance.[33]

The Breakfast Club had many functions. According to Harris, the ACDP assigned tasks to different components and their designated participants, and the Breakfast Club would coordinate the resulting products, work through drafts, and help figure out what the next ACDP meeting agenda should be. To a large extent, she said, the Breakfast Club was a venue for sharing information and ideas rather than an action group. That said, she also noted that Work used the Breakfast Club for various other tasks, such as getting help with speeches and presentations and vetting them.[34] For example, the Breakfast Club vetted the speech on the Third Offset that Work gave in September 2015 to the Royal United Services Institute in the United Kingdom. The Breakfast Club also vetted some of Work's congressional briefings, and he used the group for staff augmentation or to serve as what Harris described as a "special think tank." Later, during the transition to Patrick M. Shanahan, who was Work's replacement, the Breakfast Club produced documents to help Shanahan and other new officials get up to speed. Among other things, the Breakfast Club wrote a history of the ACDP.

---

[30] Harris, 2019. These individuals were characterized as the "direct agents" of higher-ranking DoD officials.

[31] Bajraktari, 2019.

[32] Harris, 2019.

[33] Harris, 2019.

[34] Harris, 2019.

As was the case with the ACDP, the views of the Breakfast Club participants evolved. Originally, according to Bajraktari, roughly one-third of the participants believed in the value of the Third Offset, one-third were there because they were ordered to be, and one-third were there because, although they were skeptics, they were interested in learning more.[35] Bajraktari credits Work with motivating those who were already convinced of the value of the Third Offset to make considerable progress and accomplish a lot, notwithstanding the fact that they were initially in the minority. Work, Bajraktari said, knew how to motivate the staff, and people knew that when Breakfast Club participants spoke, they were doing so with the backing of Work himself. By fostering an "excellent culture," moreover, Work both strengthened the existing believers and attracted more. He also found ways to incentivize initiatives that were tied directly to the Third Offset. For example, according to Bajraktari, Work presented DoD award coins to members of the J-8 who held Third Offset wargames.[36]

DARPA, Defense Innovation Unit—Experimental, and the Strategic Capabilities Office

While the ACDP and the Breakfast Club provided support to Third Offset goals, DARPA (which obviously predates the Third Offset by many decades), DIUx, and SCO played a central role in investing in technological innovation. To paraphrase Work, the idea was for the three organizations to fit together along a continuum.[37] DARPA focused on the most-advanced technologies and was, according to Work, "looking out on the 20-year horizon and beyond for what technologies might empower military operations in the future."[38] DIUx had a nearer-term focus on the particular mission of engaging with the commercial technology sector, especially in Silicon Valley. SCO,

[35] Bajraktari, 2019.

[36] Bajraktari, 2019.

[37] Selva and Work, 2016.

[38] Selva and Work, 2016.

in contrast, was "looking at taking current capabilities, mixing them in different ways, and actually doing demonstrations of capabilities that could emerge in the next five to ten years, but are not here today because of the way we choose to organize and mix weapon systems."[39] In the next subsections, we take a brief look at DIUx and SCO, leaving aside DARPA because it was, relatively speaking, less a reflection of the Third Offset and its particular priorities than were these newer entities.

Defense Innovation Unit—Experimental

Among the many initiatives generated by the Third Offset, one stands out as particularly emblematic: DIUx (which is, as of December 2020, more commonly referred to as simply *DIU*).[40] This is because it represents the fruit of Work's and others' thinking regarding the relationship between the U.S. government and technology. Specifically, a pillar of Third Offset thinking is the observation that, during the First and Second Offsets, the U.S. government—and, more particularly, DoD—largely drove technological innovation through its investments, in collaboration with government labs and large established defense contractors, such as Boeing and Lockheed Martin. In the second decade of the 21st century, however, innovation was coming from industry, and, above all, from the new technology firms associated with Silicon Valley in the San Francisco Bay area. It was the conviction of the leaders of the Third Offset that DoD would have to forge an entirely new way of working with industry and, most especially, learn how to collaborate with Silicon Valley.

Many credit Ashton Carter with the initiative that would become DIUx. According to Maynard Holliday—who at the time worked as the special assistant for Frank Kendall, then the Under Secretary of Defense for Acquisition, Technology, and Logistics—Carter "understood that the script had been flipped" in terms of technological investment and innovation. Carter saw that DoD was no longer the primary

---

[39] Selva and Work, 2016.

[40] Billy Mitchell, "'No Longer an Experiment'—DIUx Becomes DIU, Permanent Pentagon Unit," FedScoop, August 9, 2018.

engine of innovation, especially when it came to technology that had a global reach and a global market.[41] In other words, Silicon Valley firms were selling products to a global market and therefore had little natural interest in bespoke projects for DoD that they would not be able to market widely. According to Holliday, in April 2015, Carter was invited to give the Drell Lecture at Stanford University, at which he first broached the idea of DIUx in public. He then asked Kendall to make the idea a reality.

The obvious inspiration for DIUx was In-Q-Tel, which the CIA had created in 1999. The CIA designed In-Q-Tel to function as a venture capital firm. It worked with technology firms and start-ups to identify and invest in promising new technologies that might prove valuable for the IC. As a result, it provided the CIA with an entirely novel way to collaborate with Silicon Valley firms in a manner the firms understood and found profitable.

Holliday recounts how Kendall tasked him with helping set up DIUx, largely because of Holliday's familiarity with Silicon Valley, where he had lived and worked. Also part of the team that Kendall formed was a veteran DoD civilian, Dale Ormand, and George Duchak. Duchak, who had served as the director of the Air Force Research Laboratory Information Directorate, would be the first director of DIUx. Former Navy Sea, Air, and Land (SEAL) Rear Admiral Brian Hendrickson was selected as Duchak's deputy. According to Holliday, the thinking behind that choice was that DIUx needed "an operator who understood operations and could vet technology."[42]

Together, they first scouted sites for DIUx and acquired office space in Mountain View, California. According to Holliday, he insisted that DIUx office spaces resembled what Silicon Valley people were familiar with, such as open, flexible workspaces, as opposed to "a Dilbert cubical farm with lots of security." The team toured the head-

[41] Maynard Holliday, telephone interview with RAND Corporation researchers about the Third Offset and DIUx, August 23, 2019.

[42] Holliday, 2019.

quarters of such firms as Facebook and Google and ended up hiring the same design firm that had worked with them.[43]

They also had to decide on a business model for DIUx. Holliday said that it had become clear from talking to people in his networks in Silicon Valley that, although the In-Q-Tel model offered certain strengths, it also had some disadvantages. According to Holliday, "What In-Q-Tel does well is be able to take classified intelligence problems and turn them into unclassified problem statements for the Valley to solve."[44] He said that DIUx "definitely wanted to emulate that." As a venture capital firm, however, In-Q-Tel, in exchange for its investment, received some portion of the companies in which it invested. Although attractive to some, this arrangement was a disincentive for others. It was decided—Holliday credits Kendall and Steve Welby with this decision—that, rather than copy In-Q-Tel and act as a venture firm, DIUx would focus on offering contracts for specific requirements that could be executed in 30 to 60 days.[45]

Holliday explained that DIUx had four "attractants" that served as an incentive for Silicon Valley firms, especially start-ups, to work with it. The first was that, unlike In-Q-Tel, DIUx was not asking for a piece of the company in which it invested (i.e., it was offering a non-dilutive capital investment). Second, DIUx could offer "fast-tracked patent review" through the U.S. Patent and Trademark Office if the technology met an acute national security need. Third, DIUx offered the ability to introduce firms to "tier-1 defense companies," which could pay top dollar to license the technology or perhaps buy the firms outright. The fourth was that DoD offered access to a panoply of virtual and actual test ranges.[46]

DIUx got off the ground with some difficulty, and, in May 2016, Carter replaced Duchak with Raj Shah, who came to DIUx from a

[43] Holliday, 2019.

[44] Holliday, 2019.

[45] Holliday, 2019.

[46] Holliday, 2019.

cybersecurity firm.[47] Shah's position was elevated to provide direct support to Carter, and Shah was given a budget and a measure of maneuver room that Duchak had not enjoyed.[48] Finally, Carter reportedly gave DIUx more of a "national focus," which translated into the opening of offices in Austin, Texas; Boston, Massachusetts; and Washington, D.C. The organization began to flourish.

According to Shah, the key goal for him and his team was to get advanced commercial technology into the hands of warfighters quickly.[49] The average time between putting out a solicitation and writing a contract was 60 days, he said, with expected delivery within six months. To be more effective, Shah explained, DIUx would initiate contracts only if they had an "end customer" who told them that something was a real need and that they were willing to spend the resources required to meet that need. "If they couldn't find a couple million [dollars] to put behind [the need], it wasn't a real priority," Shah said. The practice was to "co-invest." DIUx would put in, for example, $1 million, while the service would contribute $2 million. Shaw said that he drummed up work by meeting with all of the service chiefs; traveling to Washington, D.C., every week; and traveling to meet all of the various CCMDs and program offices. After a while, he said, "people learned to call us."[50]

None of the technology programs with which DIUx was involved could be described as providing the kind of profound paradigm shift evoked by the term *Third Offset*. The programs tended to develop useful technologies that met real requirements. For example, DIUx worked on a project commissioned by the SEALs, who wanted to protect the first person who would walk into a building to do a kill-

[47] Colin Clark and Sydney J. Freedberg, Jr., "SecDef Carter Unveils DIUX 2.0; Cans Current Leadership," *Breaking Defense*, May 11, 2016.

[48] Raj Shah, telephone interview with RAND Corporation researchers about the Third Offset, June 12, 2019. See also Clark and Freedberg, 2016. Duchak had reported to Stephen Welby, the Assistant Secretary of Defense for Research and Engineering.

[49] Further research addressing the impact of the Third Offset on DoD acquisitions and contracting processes, which is not covered in depth in this report, is warranted.

[50] Shah, 2019.

capture mission. That person tended to get wounded or killed. The SEALs were hoping to develop some sort of robust body armor, which proved impractical, so DIUx steered them to five engineers from Massachusetts Institute of Technology, who built a small drone that could go into a room and use light detection and ranging to map it rapidly and identify where people were. They came up with the tactic of using two drones. One would blow the door to a room; the second would enter the room and scan. It worked. This led to a $1.5-million contract, which was cheap by DoD standards, and, according to Shah, "within a few months we had the technology."[51] This surely was a good innovation, especially if it saved the lives of special forces operators, but the ambitions of the Third Offset demanded not only the development of new, lifesaving technologies but also the cultivation of an entirely new way of doing business within DoD.

When asked about the value of the technology that DIUx helped generate, Shah replied that, in effect, it did not matter. "In my view the projects we did were tremendous and great," he said. This was because "the real value" was not the technology itself but rather that the process of developing the technology showed DoD that they could do things differently.[52] This is not to say that the technologies being developed were of no value to DoD; it is just to say that the processes being developed alongside these technologies had the potential to add even more value over time. The new processes meant that DoD could interact with new suppliers—and do it quickly. Shah even dismissed the idea that there was one technology—or even that there were several technologies—that could amount to a new offset. For him, what mattered more were speed and the ability of DoD to absorb all of the new innovations coming from Silicon Valley and elsewhere. The drones developed for the SEALs, as an example, were not important in and of themselves. What mattered was establishing a precedent and a means of speeding up tech acquisition and absorption that would lead, presumably, to much more strategically significant future developments than a few tactical drones. Shah also insisted that DIUx would have

[51] Shah, 2019.

[52] Shah, 2019.

achieved much less had he aimed higher. The Pentagon would have rejected out of hand a multiyear, multibillion-dollar proposal. Shah said that he understood that by aiming for smaller, more-achievable projects, he could "put some wins on the board," which would serve as evidence that this new way of doing things could be effective. Over time, this strategy worked: DIUx grew, as did the size of some of the projects, and the example that it set became more compelling. Ultimately, Shah asserted, DIUx was less about developing specific technologies than it was about "transforming business practices." From this standpoint, Shah is satisfied that DIUx—and, arguably, even the Third Offset—was a success. Indeed, one news article noted that DIUx under Shah not only helped defense agencies and military branches "more rapidly acquire innovative tech" but also served as "a valuable resource in training DOD acquisition professionals how to do so on their own."[53]

Emblematic of that change, of course, was Shanahan's decision, announced in a memo dated August 3, 2018, to change the name of DIUx to *DIU* to mark the organization's transition from an experiment to a permanent organization. Secretary of Defense James Mattis reportedly said at a press conference, "There is no doubt in my mind that DIUx will not only continue to exist, it will . . . grow in its influence and its impact on the Department of Defense.[54]

The Strategic Capabilities Office

Another important Third Offset–related institution was SCO, which, in contrast to DIUx, was tasked with repurposing technologies that already existed within DoD. In fact, SCO predated the Third Offset, having been created by then–Deputy Secretary of Defense Carter in August 2012, but its mission aligned closely with Third Offset priorities. Specifically, Carter reportedly took the initiative after realizing that competition with China and Russia was going to oblige the

[53] Mitchell, 2018.

[54] Mitchell, 2018.

Pentagon to bring back dormant capabilities and create new ones.[55] SCO's specific mission was to look for relatively quick and inexpensive solutions by finding new ways to use extant technology, leaving the development of near- and far-term innovations to DIUx and, of course, DARPA. Naturally, the advent of the Third Offset gave SCO a boost. According to one source, its budget grew from $125 million in 2014 to $530 million in 2016.[56]

According to SCO director Will Roper, SCO had three approaches: (1) taking something designed for one mission and making it do a completely different mission; (2) integrating discrete systems into broader, integrated systems that could do something that the component systems could not do on their own; and (3) altering a capability by adding commercial technology.[57] There are several examples of SCO projects cited in open sources. One is the so-called arsenal plane, which amounted to converting B-52s into flying magazines by stuffing them with the latest sensors and weapons and making them fully linked to fifth-generation aircraft.[58] Other examples include flying and underwater swarming micro drones and hypervelocity projectiles that could be fired by artillery that is already in the U.S. military's inventory.[59] The high-velocity projectiles project, it should be noted, reflects a Third Offset concern with finding cheaper countermeasures against relatively inexpensive weapons. The projectiles are intended to be used against incoming missiles; anti-missile defenses historically have involved firing expensive missiles to intercept much cheaper incoming weapons, which tends to give the attacker an inherent advantage.

Perhaps the most cited example of SCO's achievements, however, was the transformation of the SM-6 surface-to-air missile into an anti-ship missile. As one article noted, "repurposing defensive missiles as

---

[55] Cheryl Pellerin, "DoD Strategic Capabilities Office Gives Deployed Military Systems New Tricks," U.S. Department of Defense, April 4, 2016.

[56] Colin Clark, "Robot Boats, Smart Guns & Super B-52s: Carter's Strategic Capabilities Office," *Breaking Defense*, February 5, 2016.

[57] Pellerin, 2016.

[58] Clark, 2016.

[59] Clark, 2016.

offensive ones . . . reflects a Pentagon push to make old weapons do new tricks for a minimum added cost."[60] The adaptation, moreover, was representative of the military's new strategic focus under the Third Offset: "After the Soviet Union fell in 1991, the US Navy refocused from fighting hostile fleets to striking land targets."[61] Air defenses also became a higher priority. The result was that "destroyers and cruisers increasingly filled their missile tubes with Tomahawk Land Attack Missiles and defensive Standard Missiles,"[62] which left less room for anti-ship weapons. Those that remained were all variants of the Harpoon, which has less range than newer Russian and Chinese systems. Therefore, it had become apparent, now that the Navy was once more focusing on dealing with the threats posed by the fleets of rival powers, that there was a requirement for a better anti-ship capability. Adapting the SM-6 met that requirement, reportedly at a modest cost.

SCO also reflected what seems to have been a strategic choice on the part of Carter—and a choice that squared with Work's own thinking about how to achieve Third Offset objectives. Specifically, given the relative novelty of Third Offset thinking, or even just the shift in focus to great-power competition, there was a sense that one needed to aim relatively low, in hopes of small wins, rather than be more ambitious. One article was explicit in couching everything in budgetary terms: At the end of the Obama administration, it was simply more plausible to make relatively small budget requests than to ask for major programs and attempt major reforms. "That's why Defense Secretary Ash Carter's Strategic Capabilities Office focuses on relatively small investments that get more use out of existing assets," one article stated.[63] It is also why "the imperative to counter China and Russia is not

---

[60] Sydney J. Freedberg, Jr., "Anti-Aircraft Missile Sinks Ship: Navy SM-6," *Breaking Defense*, March 7, 2016b.

[61] Freedberg, 2016b.

[62] Freedberg, 2016b.

[63] Sydney J. Freedberg, Jr., "DepSecDef Work Details 2017 Budget: Offset Just Beginning Exclusive," *Breaking Defense*, February 9, 2016a.

launching any major modernization programs, but rather upgrading and upgunning existing systems."[64]

One indicator of SCO's success is Carter's decision in November 2016 to make it a "permanent structure."[65] Like DIUx, what had been an experiment only a few years before had become part of the established structure of the Pentagon. Tellingly, in 2019, there was an effort to move SCO under DARPA, reportedly instigated by Under Secretary of Defense for Research and Engineering Mike Griffin, who overseas both SCO and DARPA.[66] The move was resisted by several senior leaders in DoD and the services, on the grounds that the move would bury SCO under layers of oversight and prevent it from acting quickly.[67] Congress ultimately rejected the move and acted to protect SCO by placing it under the control of Deputy Secretary of Defense David Norquist.

Cost Assessment and Program Evaluation and the Third Offset

Another organization that supported the Third Offset was the Office of the Secretary of Defense—Cost Assessment and Program Evaluation (CAPE). Although CAPE predated the Third Offset, it supported the Third Offset by helping Work realize his agenda. The most concrete example of CAPE's support for the Third Offset can be seen in the SPRs (colloquially pronounced "spears"). Elaine Simmons, CAPE's deputy director at the time, explained that, prior to Work's arrival, CAPE balanced between capabilities and institutional issues, such as

---

[64] Freedberg, 2016a.

[65] Dave Majumdar, "The Pentagon's Strategic Capabilities Office (SCO) Takes Center Stage," *National Interest*, November 17, 2016.

[66] Colin Clark, "Top DoD Official Shank Resigns; SCO Moving to DARPA," *Breaking Defense*, June 17, 2019.

[67] Theresa Hitchens, "Hill to Griffin: No Moving the SCO; Shifts It to DepSecDef Norquist," *Breaking Defense*, December 17, 2019.

sexual harassment.[68] Work, however, wanted to focus on capabilities. Thus, for example, according to Simmons, CAPE wanted to commission the RAND Corporation to do a study on suicide among service members. Work said no because he wanted CAPE to focus on the development of new capabilities.[69]

Instead, CAPE conducted SPRs that had to do with exploring ways to make the force more capable, especially, but not exclusively, with regard to countering A2/AD or finding ways to create no-man's-lands—the idea being that, if the U.S. military cannot enter a specific area, it could at least block up the adversary as well. A lot of this work had to do with "connecting grids," or bridging sensors with capabilities across different domains. Simmons said that A2/AD made multidomain operations (MDO) a necessity. As she explained it, the Army might insist that it could find its own targets. The reality of A2/AD, however, was that the Army would not be able to get close enough with its ISR assets. Therefore, the Army needed someone else to do it, and the Army also had to worry about its connectivity with that "someone else." Ultimately, CAPE's goal was to present programmatic options—i.e., clear choices to make with respect to investing in specific technologies or acquiring certain systems.

CAPE also was used as a vehicle for supporting Work's push to promote wargaming. According to Simmons, CAPE, as one of the wargaming "Quad Chairs," helped manage the fund that Work created to promote wargames and run the "beauty contest" that selected the best wargame proposals, with some preference going to ones that aligned with Third Offset priorities (for additional context, see the discussion of the Quad Chairs in the next section).

---

[68] Elaine Simmons, interview with RAND Corporation researchers, October 29, 2019.

[69] We focus on the role of CAPE here, and not ONA or OSD/P, to reflect the fact that Work relied primarily on CAPE for capabilities analysis in support of the Third Offset.

Wargames and the Third Offset

In addition to the organizations that we have noted, an increased investment in wargaming was one of the hallmarks of the Third Offset. Wargaming was one of the LOEs that was identified as part of the DII back in 2014; the utility of wargaming was promoted subsequently by the ACDP. More specifically, the Third Offset invested more heavily in wargaming while (a) finding the means to enable the larger community to benefit from them and (b) shaping them to better respond to the needs of the Third Offset. Work explained his interest in wargaming in a memo on the subject dated February 9, 2015, in which he expressed concern that DoD's "ability to test concepts, capabilities, and plans using simulation and other techniques—otherwise known as wargaming—[had] atrophied."[70] According to Simmons, Work was interested in promoting wargaming of all kinds, not just wargames related to the Third Offset.[71] That said, he also saw wargames as essential for the success of the DII because he perceived that they promoted the kind of innovation that was at its heart.

Selva, in a speech delivered October 19, 2017, to the 2017 Military Operations Research Society Wargaming Special Workshop, said that he and Work realized in 2015 that "in most of our war games inside the DoD . . . we were not recording for general consumption the results of those games."[72] No one shared, and DoD treated the game results as if they needed to be closely held, like sensitive compartmented information, for no better reason than it was accepted practice to do so. Selva recounted that he and Work stripped away some of the wargaming money held by the services and "pulled it into a corporate account." They then proposed that anyone who wanted to use the money needed to do two things: first, post the preliminary findings to the wargaming archive that the Third Offset was creating, and second,

[70] Robert O. Work, U.S. Deputy Secretary of Defense, "Wargaming and Innovation," memorandum to Pentagon leadership, Washington, D.C., February 9, 2015a, p. 1.

[71] Simmons, 2019.

[72] Paul J. Selva, "Keynote Address to the Military Operations Research Society (MORS) 2017 Wargaming Special Workshop," Alexandria, Va., October 19, 2017, p. 4.

formally request the money by submitting a request for a grant.[73] The idea behind the grant was that it would enable the J-7 Directorate for Joint Force Development to prioritize. Selva declared the effort to be successful, because there was now a "fairly robust set of war gaming conclusions" in the archive, and "no shortage of applications for money for war game initiatives."[74]

Work also created a Defense Wargaming Alignment Group that was run by four "Quad Chairs": the Office of the Under Secretary of Defense for Policy (OSD/P), CAPE, ONA, and the Joint Staff. Among other things, they managed the process of determining which proposals would receive wargame incentive funds. At an action officer level, representatives from the Quad Chairs would create a proposed list of which games to fund each quarter. This list would be reviewed at a two-star level (e.g., Deputy Assistant Secretary of Defense Mara Karlin representing OSD/P).

Work called for a three-tiered effort that focused on three time horizons: near term (0–5 years), middle term (5–15 years), and far term (beyond 15 years). The near-term effort would "focus on the execution and improvement of current operational plans and the reinvigoration of Joint combined-arms expertise."[75] Work wanted CCMDs, the services, the Joint Staff, and OSD/P to lead it. The mid-term effort would focus on "the development of new capabilities as well as operational and organizational concepts . . . with an eye toward incorporating innovative approaches or technologies into the future force and identifying potential portfolio offsets."[76] This effort would be led by the Joint Staff, "with significant participation from Policy, [CAPE], [CCMDs], and the Military Departments." As for wargaming in the far term, the idea was to "assess the operational impacts of technol-

---

73 Selva, 2017, p. 4.

74 Selva, 2017, p. 4.

75 Work, 2015a.

76 Work, 2015a.

ogy trends, future challenges, and military competitions."[77] This effort would be led by ONA.

The LRRDPP

Although we do not address the work on the LRRDPP in detail in this report, because of the classification of its activities, we briefly note it because it was significant for its technological focus and its centrality to the broader concept of the Third Offset. It was the part of the Third Offset that was the most directly involved in the cultivation of high-end technologies that might help the United States once more widen its technological lead over potential adversaries. Indeed, the LRRDPP arguably represented the essence of the Third Offset and was even an inspiration for it, if one understands the Third Offset in its first, literal meaning. This is evident in Work's characterization of a 1970s iteration of the LRRDPP as a driving force behind the Second Offset.[78] Work also repeatedly stated that he considered the term *Third Offset* to be shorthand for the LRRDPP.[79]

Winding Down

The Third Offset survived the transition to the administration of President Donald J. Trump, who took office in January 2017; the administration was arguably more attuned to the idea of treating China as a competitor than the Obama administration had been.[80] However, the change in personnel brought important shifts in focus and interest. Perhaps just as importantly, many of the ideas that had been somewhat heretical back in 2014 had become orthodoxy. Having opened the door

---

[77] Work, 2015a.

[78] Work, 2019a.

[79] Work, 2019a. Classification issues limit our discussion of the LRRDPP in this publication.

[80] For example, see Nick Corasaniti, Alexander Burns, and Binyamin Appelbaum, "Donald Trump Vows to Rip Up Trade Deals and Confront China," *New York Times*, June 28, 2016.

to a new way of thinking, the upstart insurgency that Work led had become less essential going forward.

Perhaps the first important change was the transition from Secretary of Defense Carter to Mattis. Although Mattis apparently did not disagree with the Third Offset, he had a different focus, and he did not believe that the Third Offset agenda required the same kind of attention that Carter did, or that Work would have wanted. In a speech that Selva gave on October 19, 2017, someone asked him whether the Third Offset still existed. Selva replied yes, but he then cited Mattis, who had told him that the Third Offset was not a strategy. Rather, it was a method to get at new capabilities that would allow new strategies to be built. To this, Selva had said, apparently half-jokingly, "Oh, dammit, we're going to have to go home and do some work."[81] The implication seemed to be that Mattis agreed with the precepts of the Third Offset but had downgraded it to something less than the leadership-driven movement that it had once been. In an interview with RAND researchers, Selva embraced aspects of the Third Offset while being dismissive of its overall significance. For example, he noted that Mattis boiled down much of the Third Offset to getting information to the right people at the right time, something Selva believed made a lot of sense. Mattis would embrace and promote that idea while dropping the Third Offset label.[82]

A more significant change occurred when Work himself left the job of deputy secretary in July 2017, to be replaced in the position by Shanahan. Shanahan was much less interested in the Third Offset, or, at the very least, he did not see the point of putting his own weight behind it in the way that Work had done. Many also have stressed the difference in management styles between the two men. Harris, for example, described Work as the start-up founder and disrupter.[83] In order for Work's goals to be achieved, "the changes would have to be so disruptive as to change the business model." In contrast, Shanahan, a former Boeing executive, was interested in institutionalizing and rou-

81 Selva, 2017.

82 Selva, 2017.

83 Harris, 2019.

tinizing operations. According to Harris, Work was the right guy at the right time, but so was Shanahan, and the Third Offset would have made less of an impact had the two men served in the reverse order.

In any case, without Work at the helm, the organizational constructs that had bolstered the Third Offset lost their momentum. According to Harris, Shanahan chaired the ACDP only once—on that occasion, the body briefed him on Project Maven and why it was believed that AI should be a focus. (Harris believes that the resulting set of conversations led to the creation of the Joint AI Center, and that one could therefore "draw a direct line from the ACDP to the Joint AI Center.") According to Grant, the last ACDP meeting was in June or July 2018. Harris commented that, once it became clear that Shanahan did not want the ACDP to continue, Grant's replacement, Matt Van Konynenburg, communicated to the Breakfast Club that it, too, was coming to a close.[84]

The 2018 National Defense Strategy

The diminishing interest and involvement of Shanahan and others in the ACDP and the Breakfast Club did not mean that they had abandoned the ideas at the heart of the Third Offset. On the contrary, it reflected a sense that, because many of the aims of the Third Offset were reflected in the NDS, there was less of a need for disruption or for backing organizations that were designed to make an end run around the Pentagon bureaucracy. By opening the door to the principles enshrined in the NDS, the Third Offset inadvertently led to its own demise.

It is clear from the unclassified summary of the NDS that, at the very least, DoD's leadership had arrived at the same conclusions as those of the principal advocates of the Third Offset. For example, the summary leads with the assertion that "we are emerging from a period of strategic atrophy, aware that our competitive military advantage has been eroding."[85] More boldly, it states: "Inter-state strategic competition, not terrorism, is now the primary concern in U.S. national secu-

---

[84] Harris, 2019.

[85] Mattis, 2018, p. 1.

rity." It further asserts: "China is a strategic *competitor* using predatory economics to intimidate its neighbors while militarizing features in the South China Sea," while "Russia has violated the borders of nearby nations and pursues veto power over the economic, diplomatic, and security decisions of its neighbors."[86] Back when Work was honing his ideas at CNAS, language of this sort (e.g., "competitor") would not have been accepted.

The NDS also speaks of technology in terms that are distinctly reminiscent of the Third Offset.[87] First, the very fact that the NDS places "rapid technological advancements" front and center suggests that it shares with advocates of the Third Offset a belief in the importance of facilitating cooperation between DoD and industry. Second, and relatedly, the NDS counts AI and autonomy among several technologies of particular importance. It also acknowledges that the private sector is driving a considerable portion of innovation, noting that "maintaining the Department's technological advantage will require changes to industry culture, investment sources, and protection across the National Security Innovation Base."[88]

The perception of key players in the Third Offset is that the NDS serves as proof of their success. Work views the NDS as essentially the Third Offset. He also sees such new initiatives as MDO as evidence of the Third Offset's influence. MDO, after all, is the "fruit" of having to think about peer competitors, which "tends to brace the mind." That, and the fact that some of the solutions that MDO offers are "Third Offset-y," seems to Work as indicative of the Third Offset's influence.[89] Even the use of the word "competitor" is something that Third Offset activists see as proof of their impact. Bajraktari, for example, asserted that he thought it was Carter and Work who had planted the "seeds"

---

[86] Mattis, 2018, p. 1; emphasis added.

[87] Although this shift was undoubtedly influenced by other events as well, the imprint of the Third Offset on the NDS is clear.

[88] Mattis, 2018, p. 3.

[89] Work, 2019a.

of the phrasing about "competition."[90] The phrase gained currency in large part because of the influence of the Third Offset, becoming—independent of any particular administration—a core tenet of national defense strategy. Harris similarly insisted that the NDS proved that the ACDP had influenced the increasingly widespread conviction that concepts like great-power competition and interest in specific technologies were strategic imperatives. "Had it not been for the Third Offset, I don't think the NDS would look like it does," Harris said.[91] Indeed, for Harris, the NDS meant that the ACDP had worked itself out of a job.

Elbridge Colby, the former Deputy Assistant Secretary of Defense for Strategy and Force Development, concurs, arguing that Third Offset proponents deserve credit for the creation of the NDS and that much of the Third Offset found its way into the NDS.[92] There are differences, of course. Colby notes, for example, that the NDS is not as technologically focused as the Third Offset and that it has a broader vision than the Third Offset. That said, DoD "could not have had the NDS without the Third Offset," because the NDS built on Third Offset analyses and "counter-power projection," and the Third Offset provided an "intellectual baseline" for the NDS.[93] Ultimately, according to Colby, the NDS demonstrates "total evolutionary continuity with Work's endeavors."[94]

---

[90] Bajraktari, 2019.

[91] Harris, 2019.

[92] Elbridge Colby, interview with RAND Corporation researchers about Colby's experiences while working on the Third Offset, Arlington, Va., January 10, 2020.

[93] Colby, 2020.

[94] Colby, 2020.

CHAPTER FOUR

Conclusion

We opened this history by arguing that the Third Offset can be interpreted in different ways. At the most literal level, the Third Offset refers to an initiative intended to replicate the so-called First and Second Offsets, in which the U.S. military exploited technology to offset certain specific advantages enjoyed by the Soviet Union. The use of the term *Third Offset* conveyed confidence that technology could once again enable the United States to achieve a technological advantage over its adversaries, although this time the objective was not to offset the Soviets' conventional advantages but rather to respond to specific challenges that are particular to fighting China and Russia and their advanced A2/AD capabilities.[1]

At another level, the Third Offset referred more loosely to a set of ideas. One of these ideas was the conviction that China and Russia—but especially China—were, in fact, strategic competitors and needed to be treated as such, which at times ran counter to policy orthodoxy since the end of the Cold War. The corollary to this idea was the conviction that the United States needed to develop a strategy for competing with China and Russia. This meant, among other things, refocusing the military on the kind of military capabilities required to confront peer adversaries, something that parts of DoD had not been doing for

1 It is premature to assess whether this goal will eventually be achieved.

at least a decade.[2] In particular, there was concern with countering China's and Russia's A2/AD technologies.

Another idea pertained to the relationship between DoD and industry, as well as DoD's diminished role in driving innovation, at least when compared with the 1950s or the 1970s and 1980s. The Third Offset featured a drive to find new ways to cultivate technological innovations and interact with the commercial world, including Silicon Valley. Relatedly, the Third Offset encompassed ideas about how DoD had to change how it did business, especially in relation to cultivating and acquiring new technologies, absorbing innovations, and developing entirely new operating concepts to make use of them.

One can argue that the Third Offset did not achieve what it initially set out to accomplish. At least during its brief lifespan, it did not result in a set of capabilities that offset Chinese and Russian capabilities. It also is far from clear whether any of the technologies currently in development that owe their genesis, at least in part, to the Third Offset will ever result in game-changing capabilities of the sort associated with the Second Offset. This narrow view of the impact of the Third Offset might suggest that it accomplished little more than RMA, AAN, Transformation, or any of the big technology-related enthusiasms with which the Pentagon was seized in the 1990s and early 2000s.

If one thinks of the Third Offset and its impact more broadly, to encompass the many ideas that it promoted—from the return to great-power competition to the need to shake up DoD business practices[3]—and the extent to which these ideas shaped the NDS, the picture becomes rather different. One cannot help but be impressed by the impact of the Third Offset—and the impact of its proponents, most notably Work. It is impossible, of course, to show a direct link between the major themes of the 2018 NDS and the efforts of Work and others in conceptualizing and promoting the Third Offset. It is

[2] It could be argued, of course, that Russia and China have been designated as strategic competitors to provide justification for spending on new technological capabilities. For the purposes of this report, however, we do not assess the motivations behind the characterization of Russia and China as strategic competitors.

[3] This shift might have occurred even in the absence of the Third Offset, but we argue that the Third Offset played a significant role in facilitating this shift.

clear that by 2014, several leaders, including Winnefeld, Carter, and O'Sullivan, were tracking in a similar direction. Yet Work gave these ideas a focus and a direction that hitherto had been lacking, which enabled the Third Offset to open the door to a new way of thinking about national security and defense strategy. O'Sullivan described Work's contribution in terms of a vision. Work, she observed, "wanted to inspire people and give them permission to think outside the lines."[4] This, she said, he achieved. Or, as Selva put it, "the Third Offset was shorthand for a lot of things that were going on, and those things are still persisting."[5] However, Selva sounded a note of concern regarding the extent to which the NDS ultimately will carry forward the ideas of the Third Offset. Because those ideas are enshrined in the NDS but not institutionalized more broadly, when the next defense secretary comes along and drafts a new NDS, they might be discarded, deliberately or not. People might forget.

Even so, the Third Offset and the changes that it produced inside DoD serve as a guide for how to bring about change in large governmental institutions. The history of the Third Offset showcases the importance of positive, inspired leadership, as exhibited by Bob Work, Stephanie O'Sullivan, and others who understood the problem at hand, provided a vision for its solution, and then led a team of committed public servants as they worked toward that goal. That story of organizational change and leadership represents one of the main takeaways of the history of the Third Offset.

---

[4] O'Sullivan, 2019.

[5] Selva, 2019.

References

Arthur D. Simons Center for Interagency Cooperation, "Interagency Space Center Makes Strides," March 17, 2016a. As of November 4, 2020: http://thesimonscenter.org/ia-space-center-makes-strides/

———, "JICSpOC Creates More Seamless Coordination Across Government," September 30, 2016b. As of November 4, 2020: https://thesimonscenter.org/jicspoc-creates-more-seamless-coordination-across-government/

Bacevich, A. J., *The Pentomic Era: The U.S. Army Between Korea and Vietnam*, Washington, D.C.: National Defense University Press, 1986.

Bacevich, Andrew J., "'Splendid Little War': America's Persian Gulf Adventure Ten Years On," in Andrew J. Bacevich and Efraim Inbar, eds., *The Gulf War of 1991 Reconsidered*, Abingdon, UK: Routledge, 2003, pp. 149–166.

Bader, Jeffrey A., *Obama and China's Rise: An Insider's Account of America's Asia Strategy*, Washington, D.C.: Brookings Institution Press, 2012.

Bajraktari, Ylli, interview with RAND Corporation researchers about Bajraktari's experiences while working on the Third Offset, Arlington, Va., July 17, 2019.

Biddle, Stephen, "The Gulf War Debate Redux: Why Skill and Technology Are the Right Answer," *International Security*, Vol. 22, No. 2, Fall 1997, pp. 163–174.

———, *Military Power: Explaining Victory and Defeat in Modern Battle*, Princeton, N.J.: Princeton University Press, 2004.

Bitzinger, Richard A., *Assessing the Conventional Balance in Europe, 1945–1975*, Santa Monica, Calif.: RAND Corporation, N-2859-FF/RC, 1989. As of November 3, 2020: https://www.rand.org/pubs/notes/N2859.html

Brands, Hal, *Making the Unipolar Moment: U.S. Foreign Policy and the Rise of the Post–Cold War Order*, Ithaca, N.Y.: Cornell University Press, 2016.

———, *American Grand Strategy in the Age of Trump*, Washington, D.C.: Brookings Institution Press, 2018.

Brown, Harold, *Department of Defense Annual Report Fiscal Year 1982*, Washington, D.C.: U.S. Department of Defense, January 19, 1981. As of December 9, 2020:
https://history.defense.gov/Historical-Sources/Secretary-of-Defense-Annual-Reports/

Bumiller, Elisabeth, and Michael Wines, "Test of Stealth Fighter Clouds Gates Visit to China," *New York Times*, January 11, 2011.

Chollet, Derek, *The Long Game: How Obama Defied Washington and Redefined America's Role in the World*, New York: PublicAffairs, 2016.

Chubb, Andrew, "Xi Jinping and China's Maritime Policy," Brookings Institution, January 22, 2019. As of November 4, 2020:
https://www.brookings.edu/articles/xi-jinping-and-chinas-maritime-policy/

Clark, Colin, "Robot Boats, Smart Guns & Super B-52s: Carter's Strategic Capabilities Office," *Breaking Defense*, February 5, 2016. As of November 4, 2020:
https://breakingdefense.com/2016/02/carters-strategic-capabilities-office-arsenal-plane-missile-defense-gun/

———, "Top DoD Official Shank Resigns; SCO Moving to DARPA," *Breaking Defense*, June 17, 2019. As of November 4, 2020:
https://breakingdefense.com/2019/06/top-dod-official-shank-resigns-sco-moving-to-darpa/

Clark, Colin, and Sydney J. Freedberg, Jr., "SecDef Carter Unveils DIUX 2.0; Cans Current Leadership," *Breaking Defense*, May 11, 2016. As of November 4, 2020:
https://breakingdefense.com/2016/05/secdef-carter-unveils-diux-2-0-cans-current-leadership/

Colby, Elbridge, interview with RAND Corporation researchers about Colby's experiences while working on the Third Offset, Arlington, Va., January 10, 2020.

Corasaniti, Nick, Alexander Burns, and Binyamin Appelbaum, "Donald Trump Vows to Rip Up Trade Deals and Confront China," *New York Times*, June 28, 2016.

Defense Science Board, *Summer Study on Autonomy*, Washington, D.C.: U.S. Department of Defense, June 2016. As of December 8, 2020:
https://www.hsdl.org/?abstract&did=794641

DoD—*See* U.S. Department of Defense.

Drew, Christopher, "Victory for Obama over Military Lobby," *New York Times*, October 28, 2009.

DSB—*See* Defense Science Board.

Ehrhard, Thomas P., and Robert O. Work, *The Unmanned Combat Air System Carrier Demonstration Program: A New Dawn for Naval Aviation?* Washington, D.C.: Center for Strategic and Budgetary Assessments, May 10, 2007. As of November 4, 2020:
https://csbaonline.org/uploads/documents/2007.05.10-The-Unmanned-Combat-Air-System-Carrier-Demonstration-Program.pdf

———, *Range, Persistence, Stealth, and Networking: The Case for a Carrier-Based Unmanned Combat Air System*, Washington, D.C.: Center for Strategic and Budgetary Assessments, 2008. As of November 4, 2020:
https://csbaonline.org/research/publications/range-persistence-stealth-and-networking-the-case-for-a-carr-ier-based-unma/publication/1

Fallows, James, *National Defense*, New York: Random House, 1981.

Fischer, Eric A., Edward C. Liu, John W. Rollins, and Catherine A. Theohary, *The 2013 Cybersecurity Executive Order: Overview and Considerations for Congress*, Washington, D.C.: Congressional Research Service, R42984, December 15, 2014. As of November 4, 2020:
https://fas.org/sgp/crs/misc/R42984.pdf

Freedberg, Sydney J., Jr., "DepSecDef Work Details 2017 Budget: Offset Just Beginning Exclusive," *Breaking Defense*, February 9, 2016a. As of May 28, 2020:
https://breakingdefense.com/2016/02/high-tech-seed-corn-for-next-president-bob-work-on-2017-budget/

———, "Anti-Aircraft Missile Sinks Ship: Navy SM-6," *Breaking Defense*, March 7, 2016b. As of May 28, 2020:
https://breakingdefense.com/2016/03/anti-aircraft-missile-sinks-ship-navy-sm-6/

Friedman, Uri, "The New Concept Everyone in Washington Is Talking About," *The Atlantic*, August 6, 2019.

Gaddis, John Lewis, *Strategies of Containment: A Critical Appraisal of American National Security Policy During the Cold War*, New York: Oxford University Press, 1982.

Gates, Robert M., *Duty: Memoirs of a Secretary at War*, New York: Alfred A. Knopf, 2014.

Gertz, Bill, "Report: China's Military Is Growing Super Powerful by Stealing America's Defense Secrets (Like the F-35)," *National Interest*, December 8, 2016.

Gessert, Robert A., "The AirLand Battle and NATO's New Doctrinal Debate," *RUSI Journal*, Vol. 129, No. 2, 1984, pp. 52–60.

Gleason, S. Everett, "Memorandum of Discussion at the 230th Meeting of the National Security Council," Washington, D.C., January 5, 1955.

Goldberg, Jeffrey, "The Obama Doctrine," *The Atlantic*, April 2016.

Gorman, Siobhan, August Cole, and Yochi Dreazen, "Computer Spies Breach Fighter-Jet Project," *Wall Street Journal*, April 21, 2009.

Grant, Greg, interview with RAND Corporation researchers about Grant's experiences while working on the Third Offset, telephone, September 27, 2019.

Hagel, Chuck, "Reagan National Defense Forum Keynote," speech, Simi Valley, Calif., November 15, 2014a.

———, U.S. Secretary of Defense, "The Defense Innovation Initiative," memorandum to U.S. Department of Defense staff, Washington, D.C., November 15, 2014b.

Hamm, Manfred R., "The AirLand Battle Doctrine: NATO Strategy and Arms Control in Europe," *Comparative Strategy*, Vol. 7, No. 3, 1988, pp. 183–211.

Harris, Kathryn, telephone interview with RAND Corporation researchers about Harris's experiences while working on the Third Offset, June 19, 2019.

Heginbotham, Eric, Michael Nixon, Forrest E. Morgan, Jacob L. Heim, Jeff Hagen, Sheng Li, Jeffrey Engstrom, Martin C. Libicki, Paul DeLuca, David A. Shlapak, David R. Frelinger, Burgess Laird, Kyle Brady, and Lyle J. Morris, *The U.S.-China Military Scorecard: Forces, Geography, and the Evolving Balance of Power, 1996–2017*, Santa Monica, Calif.: RAND Corporation, RR-392-AF, 2015. As of November 4, 2020:
https://www.rand.org/pubs/research_reports/RR392.html

Hitchens, Theresa, "Hill to Griffin: No Moving the SCO; Shifts It to DepSecDef Norquist," *Breaking Defense*, December 17, 2019. As of November 4, 2020:
https://breakingdefense.com/2019/12/
hill-to-griffin-no-moving-the-sco-shifts-it-to-depsecdef-norquist/

Holliday, Maynard, telephone interview with RAND Corporation researchers about the Third Offset and DIUx, August 23, 2019.

Howard, Michael, "The Use and Abuse of Military History," *RUSI Journal*, Vol. 107, No. 625, 1962, pp. 4–10.

Jensen, Benjamin M., "The Role of Ideas in Defense Planning: Revisiting the Revolution in Military Affairs," *Defence Studies*, Vol. 18, No. 3, July 2018, pp. 302–317.

Kagan, Frederick W., *Finding the Target: The Transformation of American Military Policy*, New York: Encounter Books, 2006.

Keller, John, "Navy Interested in New Computing and Sensor Technologies for Shipboard and Submarine Sonar," *Military & Aerospace Electronics*, July 10, 2017.

Lay, James S., Jr., *A Report to the National Security Council*, Washington, D.C.: National Security Council, NSC 162/2, October 30, 1953. As of December 10, 2020:
https://fas.org/irp/offdocs/nsc-hst/nsc-162-2.pdf

Lock-Pullan, Richard, "'An Inward Looking Time': The United States Army, 1973–1976," *Journal of Military History*, Vol. 67, No. 2, April 2003, pp. 483–511.

Londoño, Ernesto, "Pentagon: Chinese Government, Military Behind Cyberspying," *Washington Post*, May 6, 2013.

MacMillan, Margaret, *Nixon and Mao: The Week That Changed the World*, New York: Random House, 2007.

Mahnken, Thomas G., and Barry D. Watts, "What the Gulf War Can (and Cannot) Tell Us About the Future of Warfare," *International Security*, Vol. 22, No. 2, Fall 1997, pp. 151–162.

Majumdar, Dave, "The Pentagon's Strategic Capabilities Office (SCO) Takes Center Stage," *National Interest*, November 17, 2016.

Manyin, Mark E., Stephen Daggett, Ben Dolven, Susan V. Lawrence, Michael F. Martin, Ronald O'Rourke, and Bruce Vaughn, *Pivot to the Pacific? The Obama Administration's "Rebalancing" Toward Asia*, Washington, D.C.: Congressional Research Service, R42448, March 28, 2012. As of November 4, 2020:
https://fas.org/sgp/crs/natsec/R42448.pdf

Marshall, A. W., *Long-Term Competition with the Soviets: A Framework for Strategic Analysis*, Santa Monica, Calif.: RAND Corporation, R-862-PR, 1972. As of November 3, 2020:
https://www.rand.org/pubs/reports/R862.html

Marshall, A. W., and James Roche, "Strategy for Competing with the Soviets in the Military Sector of the Continuing Political-Military Competition," unpublished Department of Defense memorandum, 1976.

Martinage, Robert, *Toward a New Offset Strategy: Exploiting U.S. Long-Term Advantages to Restore U.S. Global Power Projection Capability*, Washington, D.C.: Center for Strategic and Budgetary Assessments, 2014. As of November 4, 2020:
https://csbaonline.org/research/publications/
toward-a-new-offset-strategy-exploiting-u-s-long-term-advantages-to-restore

Mattis, Jim, *Summary of the 2018 National Defense Strategy of the United States of America: Sharpening the American Military's Competitive Edge*, Washington, D.C.: U.S. Department of Defense, 2018. As of December 9, 2020:
https://dod.defense.gov/Portals/1/Documents/pubs/
2018-National-Defense-Strategy-Summary.pdf

McCarthy, Jim, interview with RAND Corporation researchers about the Third Offset, Arlington, Va., November 13, 2019.

McLeary, Paul, "Pentagon's Big AI Program, Maven, Already Hunts Data in Middle East, Africa," *Breaking Defense*, May 1, 2018. As of November 4, 2020:
https://breakingdefense.com/2018/05/
pentagons-big-ai-program-maven-already-hunts-data-in-middle-east-africa/

Mitchell, Billy, "'No Longer an Experiment'—DIUx Becomes DIU, Permanent Pentagon Unit," FedScoop, August 9, 2018. As of November 4, 2020: https://www.fedscoop.com/diu-permanent-no-longer-an-experiment/

Nakashima, Ellen, "U.S. Publicly Calls on China to Stop Commercial Cyber-Espionage, Theft of Trade Secrets," *Washington Post*, March 11, 2013.

O'Mara, Margaret, "The Church of Techno-Optimism," *New York Times*, September 28, 2019.

O'Sullivan, Stephanie, interview with RAND Corporation researchers about the Third Offset and Robert Work, Arlington, Va., September 18, 2019.

Obama, Barack, *The Audacity of Hope: Thoughts on Reclaiming the American Dream*, New York: Crown Publishing Group, Crown Publishers, 2006.

———, "A New Strategy for a New World," speech, Washington, D.C., July 15, 2008. As of November 4, 2020:
https://www.wilsoncenter.org/article/new-strategy-for-new-world

———, *National Security Strategy*, Washington, D.C.: The White House, May 2010. As of November 4, 2020:
https://obamawhitehouse.archives.gov/sites/default/files/rss_viewer/national_security_strategy.pdf

Office of Public Affairs, U.S. Department of Justice, "Chinese National Admits to Stealing Sensitive Military Program Documents from United Technologies," press release, last updated December 22, 2016. As of November 4, 2020:
https://www.justice.gov/opa/pr/chinese-national-admits-stealing-sensitive-military-program-documents-united-technologies

Office of the National Counterintelligence Executive, *Foreign Spies Stealing US Economic Secrets in Cyberspace: Report to Congress on Foreign Economic Collection and Industrial Espionage, 2009-2011*, Washington, D.C.: Office of the Director of National Intelligence, October 2011. As of December 9, 2020:
https://www.hsdl.org/?abstract&did=720057

Pellerin, Cheryl, "DoD Strategic Capabilities Office Gives Deployed Military Systems New Tricks," U.S. Department of Defense, April 4, 2016. As of November 4, 2020:
https://www.defense.gov/Explore/News/Article/Article/712938/dod-strategic-capabilities-office-gives-deployed-military-systems-new-tricks/

Perry, William J., "Desert Storm and Deterrence," *Foreign Affairs*, Vol. 70, No. 4, Fall 1991, pp. 66–82.

Romjue, John L., *From Active Defense to AirLand Battle: The Development of Army Doctrine 1973–1982*, Fort Monroe, Va.: U.S. Army Training and Doctrine Command, June 1984.

Rosenberg, David Alan, "The Origins of Overkill: Nuclear Weapons and American Strategy, 1945–1960," *International Security*, Vol. 7, No. 4, Spring 1983, pp. 3–71.

Roulo, Claudette, "DoD Seeks Next-Generation Technologies, Kendall Says," U.S. Department of Defense, October 7, 2014. As of November 4, 2020: https://www.defense.gov/Explore/News/Article/Article/603398/dod-seeks-next-generation-technologies-kendall-says/

Sanger, David E., *Confront and Conceal: Obama's Secret Wars and Surprising Use of American Power*, New York: Crown Publishing Group, Broadway Paperbacks, 2012.

Selva, Paul J., "Keynote Address to the Military Operations Research Society (MORS) 2017 Wargaming Special Workshop," Alexandria, Va., October 19, 2017.

———, telephone interview with RAND Corporation researchers about the Third Offset, November 18, 2019.

Selva, Paul J., and Robert O. Work, unpublished interview about the Third Offset, October 24, 2016.

Shah, Raj, telephone interview with RAND Corporation researchers about the Third Offset, June 12, 2019.

Shane, Scott, and Daisuke Wakabayashi, "'The Business of War': Google Employees Protest Work for the Pentagon," *New York Times*, April 4, 2018.

Shimko, Keith L., *The Iraq Wars and America's Military Revolution*, New York: Cambridge University Press, 2010.

Simmons, Elaine, interview with RAND Corporation researchers, October 29, 2019.

Taylor, Maxwell D., *The Uncertain Trumpet*, New York: Harper and Row, 1960.

TheIHMC, "Paul Kaminski: STEALTH—An Insider's Perspective," video, YouTube, January 22, 2014. As of May 29, 2020: https://www.youtube.com/watch?v=10ScidoYWOY

Tomes, Robert, "The Cold War Offset Strategy: Origins and Relevance," War on the Rocks, November 6, 2014. As of November 3, 2020: https://warontherocks.com/2014/11/the-cold-war-offset-strategy-origins-and-relevance/

Trachtenberg, Marc, *A Constructed Peace: The Making of the European Settlement, 1945–1963*, Princeton, N.J.: Princeton University Press, 1999.

Transue, J. R., *Assessment of the Weapons and Tactics Used in the October 1973 Middle East War*, Arlington, Va.: Institute for Defense Analyses, Weapons Systems Evaluation Group Report 249, October 1974.

Trauschweizer, Ingo, *The Cold War U.S. Army: Building Deterrence for Limited War*, Lawrence, Kan.: University Press of Kansas, 2008.

———, *Maxwell Taylor's Cold War: From Berlin to Vietnam*, Lexington, Ky.: University Press of Kentucky, 2019.

Turpin, Matt, interview with RAND Corporation researchers about the Third Offset, Washington, D.C., June 26, 2019.

U.S.-China Economic and Security Review Commission, "Annual Reports," webpage, undated. As of December 10, 2020:
https://www.uscc.gov/annual-reports

———, *2012 Report to Congress of the U.S.-China Economic and Security Review Commission*, Washington, D.C.: U.S. Government Printing Office, November 2012. As of November 4, 2020:
https://www.uscc.gov/sites/default/files/annual_reports/2012-Report-to-Congress.pdf

———, *2013 Report to Congress of the U.S.-China Economic and Security Review Commission*, Washington, D.C.: U.S. Government Printing Office, November 2013. As of November 4, 2020:
https://www.uscc.gov/sites/default/files/annual_reports/Complete%202013%20Annual%20Report.PDF

———, *2014 Report to Congress of the U.S.-China Economic and Security Review Commission*, Washington, D.C.: U.S. Government Printing Office, November 2014. As of December 28, 2020:
https://www.uscc.gov/sites/default/files/annual_reports/Complete%20Report.PDF

U.S. Department of Defense, *Quadrennial Defense Review Report*, Washington, D.C., February 2010. As of December 9, 2020:
https://history.defense.gov/Historical-Sources/Quadrennial-Defense-Review/

U.S. House of Representatives, *Hearings on Military Posture and H.R. 10929: Department of Defense Authorization for Appropriations for Fiscal Year 1979: Hearing Before the Committee on Armed Services, House of Representatives, Ninety-Fifth Congress, Second Session*, HASC No. 95-56, Part 3, Book 1, Washington, D.C.: U.S. Government Printing Office, 1978.

U.S. Senate, *Nominations Before the Senate Armed Services Committee, First Session, 111th Congress: Hearings Before the Committee on Armed Services*, Washington, D.C.: U.S. Government Printing Office, 2010.

VanDeMark, Brian, *Road to Disaster: A New History of America's Descent into Vietnam*, New York: HarperCollins Publishers, 2018.

Vickers, Michael G., and Robert C. Martinage, *Future Warfare 20xx Wargame Series: Lessons Learned Report*, Washington, D.C.: Center for Strategic and Budgetary Assessments, December 2001.

———, *The Revolution in War*, Washington, D.C.: Center for Strategic and Budgetary Assessments, December 2004. As of November 4, 2020: https://csbaonline.org/research/publications/the-revolution-in-war/publication/1

Westad, Odd Arne, *The Global Cold War: Third World Interventions and the Making of Our Times*, New York: Cambridge University Press, 2005.

Whetten, Lawrence, and Michael Johnson, "Military Lessons of the Yom Kippur War," *The World Today*, Vol. 30, No. 3, March 1974, pp. 101–110.

Winnefeld, James, Jr., interview with RAND Corporation researchers about the Third Offset, McLean, Va., July 15, 2019.

Wohlstetter, Albert, "Is There a Strategic Arms Race?" *Foreign Policy*, No. 15, Summer 1974, pp. 3–20.

Work, Robert O., "The Coming Naval Century," *Proceedings*, Vol. 138, No. 5, May 2012.

———, "A New Global Posture for a New Era," transcript of speech delivered to Council on Foreign Relations, Washington, D.C., September 30, 2014. As of December 9, 2020:
https://www.defense.gov/Newsroom/Speeches/Speech/Article/605614/a-new-global-posture-for-a-new-era/

———, U.S. Deputy Secretary of Defense, "Wargaming and Innovation," memorandum to Pentagon leadership, Washington, D.C., February 9, 2015a.

———, "Remarks by Defense Deputy Secretary Robert Work at the CNAS Inaugural National Security Forum," transcript, Center for a New American Security, December 14, 2015b. As of November 4, 2020:
https://www.cnas.org/publications/transcript/remarks-by-defense-deputy-secretary-robert-work-at-the-cnas-inaugural-national-security-forum

———, U.S. Deputy Secretary of Defense, "Establishment of an Algorithmic Warfare Cross-Functional Team (Project Maven)," memorandum to U.S. Department of Defense staff, Washington, D.C., April 26, 2017.

———, interview with RAND Corporation researchers about the Third Offset, Arlington, Va., June 24, 2019a.

———, email to RAND Corporation researchers about the Third Offset, September 28, 2019b.

Work, Robert O., and Shawn Brimley, *20YY: Preparing for War in the Robotic Age*, Washington, D.C.: Center for a New American Security, January 2014. As of November 4, 2020:
https://www.cnas.org/publications/reports/20yy-preparing-for-war-in-the-robotic-age

Work, Robert O., and F. G. Hoffman, "Hitting the Beach in the 21st Century," *Proceedings*, Vol. 136, No. 11, November 2010.

Wormuth, Christine, interview with RAND Corporation researchers about the Third Offset, Arlington, Va., June 14, 2019.

CPSIA information can be obtained
at www.ICGtesting.com
Printed in the USA
LVHW051931100521
687012LV00013B/658